On the Splitting of Invariant Manifolds in Multidimensional Near-Integrable Hamiltonian Systems

Memoirs
of the
American Mathematical Society

Number 775

On the Splitting of Invariant
Manifolds in Multidimensional
Near-Integrable Hamiltonian
Systems

P. Lochak
J.-P. Marco
D. Sauzin

May 2003 • Volume 163 • Number 775 (second of 5 numbers) • ISSN 0065-9266

American Mathematical Society
Providence, Rhode Island

2000 *Mathematics Subject Classification.*
Primary 70H08, 70H09, 37J40, 37D10, 34C37.

Library of Congress Cataloging-in-Publication Data

Lochak, P. (Pierre)
  On the splitting of invariant manifolds in multidimensional near-integrable Hamiltonian systems / P. Lochak, J.-P. Marco, D. Sauzin.
    p. cm. — (Memoirs of the American Mathematical Society, ISSN 0065-9266 ; no. 775)
  "Volume 163, number 775 (second of 5 numbers)."
  Includes bibliographical references.
  ISBN 0-8218-3268-9 (alk. paper)
    1. Hamiltonian systems. 2. Invariant manifolds. I. Marco, J.-P. II. Sauzin, D., 1966– III. Title. IV. Series.

QA3.A57 no. 775
[QA614.83]
510 s—dc21
[514'.74]
                                                                                    2003040368

Memoirs of the American Mathematical Society

This journal is devoted entirely to research in pure and applied mathematics.

Subscription information. The 2003 subscription begins with volume 161 and consists of six mailings, each containing one or more numbers. Subscription prices for 2003 are $555 list, $444 institutional member. A late charge of 10% of the subscription price will be imposed on orders received from nonmembers after January 1 of the subscription year. Subscribers outside the United States and India must pay a postage surcharge of $31; subscribers in India must pay a postage surcharge of $43. Expedited delivery to destinations in North America $35; elsewhere $130. Each number may be ordered separately; *please specify number* when ordering an individual number. For prices and titles of recently released numbers, see the New Publications sections of the *Notices of the American Mathematical Society*.

Back number information. For back issues see the *AMS Catalog of Publications*.

Subscriptions and orders should be addressed to the American Mathematical Society, P. O. Box 845904, Boston, MA 02284-5904, USA. *All orders must be accompanied by payment.* Other correspondence should be addressed to 201 Charles Street, Providence, RI 02904-2294, USA.

Copying and reprinting. Individual readers of this publication, and nonprofit libraries acting for them, are permitted to make fair use of the material, such as to copy a chapter for use in teaching or research. Permission is granted to quote brief passages from this publication in reviews, provided the customary acknowledgment of the source is given.

Republication, systematic copying, or multiple reproduction of any material in this publication is permitted only under license from the American Mathematical Society. Requests for such permission should be addressed to the Acquisitions Department, American Mathematical Society, 201 Charles Street, Providence, Rhode Island 02904-2294, USA. Requests can also be made by e-mail to reprint-permission@ams.org.

Memoirs of the American Mathematical Society is published bimonthly (each volume consisting usually of more than one number) by the American Mathematical Society at 201 Charles Street, Providence, RI 02904-2294, USA. Periodicals postage paid at Providence, RI. Postmaster: Send address changes to Memoirs, American Mathematical Society, 201 Charles Street, Providence, RI 02904-2294, USA.

© 2003 by the American Mathematical Society. All rights reserved.
This publication is indexed in *Science Citation Index*®, *SciSearch*®, *Research Alert*®, *CompuMath Citation Index*®, *Current Contents*®*/Physical, Chemical & Earth Sciences*.
Printed in the United States of America.

∞ The paper used in this book is acid-free and falls within the guidelines
established to ensure permanence and durability.
Visit the AMS home page at http://www.ams.org/

10 9 8 7 6 5 4 3 2 1   08 07 06 05 04 03

Contents

Chapter 0. Introduction and Some Salient Features of the Model Hamiltonian — 1

Chapter 1. Symplectic Geometry and the Splitting of Invariant Manifolds — 9
  §1.1. Symplectic geometry: a short reminder — 10
  §1.2. Hyperbolic invariant manifolds — 13
  §1.3. Angles of Lagrangian planes: the symplectic viewpoint — 16
  §1.4. Angles of Lagrangian planes: the Euclidean viewpoint — 19
  §1.5. Symplectic isomorphisms, angles and splitting forms — 22
  §1.6. The splitting of Lagrangian submanifolds — 26
  §1.7. Lagrangian submanifolds in a cotangent bundle — 28
  §1.8. Hyperbolic tori and normally hyperbolic invariant manifolds — 31
  §1.9. The perturbative setting — 36
  §1.10. Lagrangian intersections and homoclinic trajectories — 40
  §1.11. The splitting of the invariant manifolds of hyperbolic tori — 46

Chapter 2. Estimating the Splitting Matrix Using Normal Forms — 51
  §2.1. Resonant normal forms — 53
  §2.2. Computations in the vicinity of a resonant surface — 59
  §2.3. Splitting in a perturbative setting, variance and stability — 61
  §2.4. General exponential estimates for the splitting matrix — 67
  §2.5. Persistence of tori, invariant manifolds and homoclinic trajectories — 77
  §2.6. Splitting and stability — 80

Chapter 3. The Hamilton-Jacobi Method for a Simple Resonance — 91
  §3.1. Notation and assumptions — 92
  §3.2. Formal solutions and the Hamilton-Jacobi algorithm — 93
  §3.3. Convergence and domains of analyticity — 99
  §3.4. Exponential closeness of the invariant manifolds — 106
  §3.5. Linear versus nonlinear splitting — 115
  §3.6. Some variants and possible generalizations — 120
  §3.7. A short historical tour and some concluding remarks — 126

Appendix. Invariant Tori With Vanishing or Zero Torsion — 133

Bibliography — 141

Abstract

In this text we take up the problem of the splitting of invariant manifolds in multi-dimensional Hamiltonian systems, stressing the canonical features of the problem. We first conduct a geometric study, which for a large part is not restricted to the perturbative situation of near-integrable systems. This point of view allows us to clarify some previously obscure points, in particular the symmetry and variance properties of the splitting matrix (indeed its very definition(s)) and more generally the connection with symplectic geometry. Using symplectic normal forms, we then derive local exponential upper bounds for the splitting matrix in the perturbative analytic case, under fairly general circumstances covering in particular resonances of any multiplicity. The next technical input is the introduction of a canonically invariant scheme for the computation of the splitting matrix. It is based on the familiar Hamilton-Jacobi picture and thus again is symplectically invariant from the outset. It is applied here to a standard Hamiltonian exhibiting many of the important features of the problem and allows us to explore in a unified way the question of finding lower bounds for the splitting matrix, in particular that of justifying a first order computation (the so-called Poincaré-Melnikov approximation). Although we do not specifically address the issue in this paper we mention that the problem of the splitting of the invariant manifold is well-known to be connected with the existence of a global instability in these multidimensional Hamiltonian systems and we hope the present study will ultimately help shed light on this important connection first noted and explored by V.I.Arnold.

2000 *Mathematics Subject Classification.* 70H08, 70H09, 37J40, 37D10, 34C37.

Key words and phrases. Separatrix splitting, Near-integrable Hamiltonian systems, Invariant manifolds, Arnold diffusion, Nekhoroshev Theorem.

CHAPTER 0

Introduction and Some Salient Features of the Model Hamiltonian

The search for and study of invariant manifolds is a long established and by now traditional part of dynamical system theory. In the more restricted but particularly difficult setting of canonical perturbation theory, the relevance of this approach was very much enhanced by the discovery of the connection of the splitting of invariant manifolds with a possible form of global instability in the near-integrable multidimensional Hamiltonian systems. Since this connection was established by V.Arnold in his famous 1964 note ([A1]) progress has been rather slow and sometimes confused in this difficult area. We hope the reader may get some feeling of the state of affairs by following the present text and the reading suggestions we have included all along, particularly of course about the splitting problem, which forms only a small and perhaps not even unavoidable part of the more general and geometric question of the global instability. We are not going to try to give an overview of all the trends and motivations in this introduction. We notice however that the end of this introduction can offer some clues on the splitting problem and that the closing sections of Chapters 2 and 3 (§2.6 and §3.7) are partly of expository and indeed historical nature and can hopefully be used as active surveys of some parts of the subject.

The present article started in part from the desire to clarify certain basic issues connected with the splitting of invariant manifolds in the higher dimensional case, essentially by injecting more geometry into problems which are partly analytic in essence but do benefit from the introduction of some symplectic geometry, as we hope will be apparent below. All the more because the problem of global instability ("Arnold diffusion") does indeed require a rather global geometric viewpoint, so that it seems only natural to start with geometrizing the local situation as much as possible. Moreover many salient features of this local situation are already apparent in a deceptively simple-looking problem embodied in the model Hamiltonian (∗) described towards the end of this introduction, which generalizes Arnold's original example and may serve as a guide into more general situations.

As the reader may have guessed, the present paper is not really meant to be introductory, and we do occasionally assume some familiarity with the subject on her or his part. Again the necessary information can easily be gathered from the reading of recent—or less recent—papers, to which we will point along the way. Each chapter is provided with its own more specific introduction, which is why this general introduction has been kept to a minimum. All along we have tried to stress and sometimes recall some basic and important ideas, occasionally indulging in some story telling in order to present these ideas in a perhaps more vivid way. This is in particular at the expense of "effectivity". We have made almost no effort to compute or estimate most of the constants which appear, although careful reading will reveal that this is usually painfully but easily doable. We have also at times been sketchy in the exposition of essentially routine proofs.

We now briefly sketch the content of the paper and then detail some features of the model problem. Our primary object of study is the splitting of the invariant

---

Received by the editor July 1, 2000.

manifolds attached to geometrically invariant objects in near-integrable higher dimensional Hamiltonian systems. This is not an *a priori* univocally defined notion and indeed Chapter 1 is devoted in large part to exploring the symplectic geometry which makes it possible to define and study several notions of "splitting of invariant manifolds" which are both quantitative and geometrically meaningful. In that opening chapter, the setting is rather general and most part of the discussion does not assume that the system under study is close to integrable. In fact many features of the problems (hyperbolic tori, invariant manifolds, splitting of Lagrangian manifolds etc.) are not perturbative in nature and much can be said about the geometry already at that level. Now, when specializing to the perturbative setting, the multidimensional near-integrable case involves constraints coming from the symplectic character of the problem, which of course in the 2-dimensional case amounts to area preservation. These are discussed in Chapter 2 but oddly enough, some fundamental geometric features (perturbative or not) do not seem to have yet been investigated as such in any detailed way. In particular, we will see that they immediately entail symmetry properties of the splitting matrix and also govern its variance, *i.e.* the way it transforms under canonical transformations. So the results in Chapter 1 arise from elementary (and sometimes less elementary) considerations of symplectic geometry. We have tried to make this chapter almost self-contained and have not refrained from going into details, as this hopefully provides a sound geometric basis for the subject, beyond what is used in the sequel of the paper.

In Chapter 2 we broach the subject of exponential smallness. This makes sense only in a *perturbative* and *analytic* situation, which is thus the natural setting of that section. Starting from such a near-integrable analytic Hamiltonian system, we look over phase space for homoclinic trajectories biasymptotic to invariant tori. One of the main results is a refined version of the result on the exponential smallness of the splitting matrix; in particular, instead of estimating the determinant of the matrix, we get estimates for the eigenvalues and indeed the eigenspaces, taking advantage of the fact that we are dealing with a symmetric matrix and thereby getting more refined information. The proofs rely on classical tools of perturbation theory, combining resonant normal forms to an exponentially high order with KAM-type results. But that these tools can be used to good effect in a relatively simple way comes for a large part from the geometric study in Chapter 1. In particular we have to carefully keep track of the effect of normalizing canonical transformations on the splitting matrix and related objects. The method used in this chapter we call *symplectic*, as it is essentially based on symplectic normal forms even if analyticity is a necessary condition for the validity of the results. The chapter closes with a rather detailed and historical discussion of the relationship between exponential stability of the action variables and the splitting of invariant manifolds.

The beginning of Chapter 3 is devoted to the description of a canonical algorithm for computing the splitting matrix, which to a large extent amounts to using the Hamilton-Jacobi method. Although this seems quite natural in the problem at hand, it has never appeared in the literature. Indeed some papers on higher order computations emphasize the wide applicability of their method, beyond Hamiltonian systems. It is however obvious that in the Hamiltonian case, much is lost by disregarding or at least not fully exploiting the symplectic nature of the problem. This is the case if one iterates the first order variational equation in order to get higher order terms, as in most techniques which have been recently developed. We then show that this Hamilton-Jacobi method provides a natural setting for the elaboration of what we term the *analytic* method for evaluating the splitting, stemming (in the lowest dimensional case) from the work of V.F.Lazutkin. This method remains to-date the most efficient one to discuss the validity of the first order (or indeed any finite order) approximation, a problem sometimes refered to as "justifying the Poincaré-Melnikov approximation" which we take up here. We also include a discussion on several issues, comparing the symplectic method used in Chapter 2 and the analytic one developed in Chapter 3. Chapter 3 focuses on the study of a model problem (Hamiltonian (∗) below) which possesses a kind of universal character, as discussed in Chapter 2. It serves to exhibit and exemplify many features of the general situation which we review below in this introduction, where the reader will find a short and concrete guided tour of some of the problems in the domain, based on this model Hamiltonian.

Finally the appendix shows how one can at least partly bridge the gap between isochronous and anisochronous problems, *i.e.* perturbations of linear versus nonlinear integrable systems.

At this point the reader might want to glance at the table of contents, and also take a look at the final subsection of the paper (§3.7.4) where we have gathered some general remarks which are meant to conclude rather than to introduce but which can be read in part without having gone through the paper and may give an idea of what has been or will be going on.

As mentioned above, we will devote the end of this introduction to recalling some important features of a significant model problem. It will be studied explicitly in several places, in particular in Chapter 3, but also provides a guiding thread in more general situations. So let us consider the following ℓ-dimensional Hamiltonian:

$$(*) \qquad H(q, \phi, p, I) = \omega \cdot I + \frac{1}{2}\alpha I^2 + \frac{1}{2}p^2 + \varepsilon(\cos q - 1) + \mu\varepsilon F(q, \phi, p, I).$$

The conjugate variables are given as:

$$(q, \phi) \in \mathbb{R}/2\pi\mathbb{Z} \times (\mathbb{R}/2\pi\mathbb{Z})^d, \quad (p, I) \in \mathbb{R} \times \mathbb{R}^d;$$

so $\ell = d + 1$, ε and μ are real perturbation parameters, ω and α are real d-dimensional vectors. The perturbation $F(q, \phi, p, I)$ is assumed to be analytic. Finally, the dot denotes the ordinary scalar product in $\mathbb{R}^d$, thus $\omega \cdot I = \sum_{1 \leq k \leq d} \omega_k I_k$. The nonlinear term αI^2 should be read as $\sum_{1 \leq k \leq d} \alpha_k I_k^2 = AI \cdot I$ where $A = \text{diag}(\alpha_k)$ is the diagonal matrix with entries given by the vector α describing the nonlinearity (alias twist or torsion) in the unperturbed system; when obvious from the context, we will use again expressions like αI (without a dot) as a somewhat daring notation for the vector given by componentwise multiplication.

When all the α_k are nonzero, which we refer to as the fully nonlinear or full torsion case, a shift of the origin in the actions I brings us back to the case where $\omega = 0$. In the latter form, it was introduced in [L1] (later included in [L2], §V.2) as a natural, yet nontrivial generalization of the system considered in [A1] and came to be known as the "Generalized Arnold Model". So in the case of full torsion bringing in the vector ω is simply a way to fix the origin of coordinates in I-space so as to focus attention near a torus with frequency ω. Here however, we will keep the frequency and twist vectors ω and α as parameters, with the possibility that some components might vanish or depend on the parameter ε. One should take notice of the manifold and slightly inconsistent terminology: *full torsion* (or *twist*) is the same as *fully nonlinear*, but also as *anisochronous*, and also *nondegenerate à la* Kolmogorov in a KAM setting. We will try to be locally consistent but such translations are sometimes unavoidable.

When $\mu = 0$, we thus have an integrable combination of uncoupled rotors and oscillators (corresponding to vanishing components α_k), together with a (p, q) pendulum. In the fully nonlinear case, KAM theory implies that if ω satisfies a Diophantine condition, the d-dimensional torus $I = p = q = 0$ with frequency ω is only slightly perturbed for μ small enough, and gives rise to a partially hyperbolic torus $\mathcal{T} = \mathcal{T}_\mu(\omega)$ (for simplicity we leave out from the notation the dependence on ε) in the perturbed system, with frequency ω and attached invariant ℓ-dimensional stable (+) and unstable (−) manifolds $\mathcal{W}^\pm = \mathcal{W}_\mu^\pm(\omega)$. These KAM results can be generalized in part to degenerate situations, in between the anisochronous or fully nonlinear case (each component of α is nonzero) and the isochronous situation ($\alpha = 0$), as discussed in the appendix. In any case, and returning here for simplicity to the fully nonlinear situation, the invariant manifolds to the perturbed torus are close (in an appropriate sense) to the unperturbed homoclinic manifold $\mathcal{W}_0^+(\omega) = \mathcal{W}_0^-(\omega)$ which is given simply by the direct product of the torus $I = 0$ in (I, ϕ)-space with the separatrix of the pendulum. The "splitting problem" is now to study the discrepancy of $\mathcal{W}_\mu^+(\omega)$ and $\mathcal{W}_\mu^-(\omega)$ as they split from the unperturbed $\mathcal{W}_0^+(\omega) = \mathcal{W}_0^-(\omega)$. One is also interested in studying the possible homoclinic intersections of the manifolds $\mathcal{W}_\mu^+(\omega)$ and $\mathcal{W}_\mu^-(\omega)$ and the geometric properties of these manifolds near the intersection trajectories. Of course, heteroclinic intersections, say of $\mathcal{W}_\mu^+(\omega)$ with $\mathcal{W}_\mu^-(\omega')$ can also be considered and are crucial in the study of instability; this is one of the main lessons to be drawn from [A1].

Hamiltonian (∗) first gives—in the fully nonlinear case—a model for the behaviour of a near-integrable Hamiltonian system with ℓ degrees of freedom near a *simple* resonance. More precisely, we have ℓ action variables (I, p). The total frequency map is given by $\Omega((I, p)) = (\omega(I), p) = (\omega + \alpha I, p)$ and we are interested in the vicinity of the simple resonance $p = 0$, a point which is discussed more generally in Chapter 2 below. We now split the discussion into a number of issues:

− Dimensions: We are interested in the case $\ell \geq 3$, *i.e.* $d \geq 2$. In the fully nonlinear case, $\ell = 3$ is the minimal dimension which makes Arnold diffusion possible, and $\ell = 3$ is special because $d = 2$ makes it possible to use the apparatus of continued fractions. This turns out to be crucial for certain issues which are solved only for $\ell = 3$ (see Chapter 3).

− Regularity: For most geometric reasonings C^2 data (reduced here to the function F) will do (see Chapters 1 and 2). However all exponential phenomena, in particular exponentially small splittings (see Chapters 2 and 3) and exponentially long stability times (see §2.6) are intimately connected with *analytic* data. Note that in both cases one can easily devise results in other categories (finite differentiability and Gevrey classes especially) which are usually easier but perhaps less interesting than in the analytic case. Moreover the rigidity of analytic functions (the fact that they cannot have compact support without vanishing altogether) make all constructions connected with instability much more difficult in the analytic category.

− Parameters: The idea of using *two* parameters comes from Poincaré ([P], end of vol. 2) and was taken up in [A1] (with the same names, *i.e.* ε and μ). The point is to avoid—or to tackle—a difficult singular perturbation problem. This is folklore and we content ourselves with the briefest indications. In the fully nonlinear case, ε brings in hyperbolicity, whereas μ destroys integrability. The most difficult, natural (in the context of perturbing a Liouville integrable system) and interesting case occurs when ε and μ are polynomially related, say $\mu = \varepsilon^p$ (take *e.g.* $p = 2$). The case when $\varepsilon = 1$ and μ is small has recently come to be known as *a priori unstable* (for the obvious reason that hyperbolicity is of order 1, *i.e.* is present in the unperturbed part), and is much easier to deal with in several respects. The aforementioned singular perturbation problem is precisely the object of Chapter 3 (and also Chapter 2, from a different viewpoint).

– Nonlinearity: As already mentioned, we call the case when all the components of α are nonzero constants the *fully nonlinear* or *full torsion* case. Then KAM-type theorems apply, in their—by now well understood—partially hyperbolic version. We will also be interested in the case when the twist vector $\alpha = \alpha(\varepsilon)$ depends on the perturbation, and some components may vanish together with ε. This case was first studied in particular cases and an elliptic setting (*i.e.* without the (p, q) variables) by G.Gallavotti and coworkers under the heading "twistless tori". We will refer to it as the *vanishing twist* case as the twist or torsion is *a priori* nonzero but does vanish with the perturbation. It turns out that this case is readily amenable to classical KAM techniques under very general conditions. This point is discussed in the appendix.

For a constant twist vector α, we note that if all components are non zero (full torsion) and moreover have the same sign, which we refer to as the *convex* case, stability of the action variables over exponentially long times prevails. If not, instability can develop over short times ($O(1/\mu)$). In other words, indefinite quadratic forms are never steep (see [N] or Appendix 9 in [LM]) and instability may develop over polynomially long times. Putting all this together, it is clear that the only case in which one should properly refer to Arnold diffusion (*i.e.* global instability over exponentially long times) is the fully nonlinear convex and analytic case, with μ and ε polynomially related. If the Hamiltonian is not fully nonlinear and convex, instability of the action variables will generically develop over polynomial (with respect to $1/\mu$) times, whereas if μ is assumed to be exponentially small with respect to ε, one finds an interesting toy problem, which however just avoids some difficult issues discussed here in Chapter 3. This last case, with μ "much" smaller than ε is actually not so different from the *a priori* unstable case ($\varepsilon = 1$, $\mu << 1$), where μ is also "much" smaller than ε; in this last case again the instability time is generically polynomial in μ (the only remaining parameter).

Returning to the various possibilities for the nonlinearity, we find at the extreme opposite of the spectrum the *torsionfree* case, when $\alpha = 0$. The only nonlinear term which remains in the integrable ($\mu = 0$) problem is then the pendulum (p, q) term.

In between the full torsion and torsionfree cases, we find a range of *mixed linear-nonlinear* (or just mixed) cases, which seem to have hardly been investigated, although some naturally arise. We will briefly return to them below. For this and other reasons, we introduce a more detailed notation for the variables: write $\omega = (\omega_1, \omega_2)$, with ω_1 an n-vector ($0 \leq n \leq d$) and ω_2 a m-vector ($m + n = d$). Correspondingly, we write $\alpha = (\alpha_1, \alpha_2)$ and $I = (I_1, I_2)$. Then the mixed cases correspond to $\alpha_2 = 0$ and all components of α_1 nonzero ($0 < m < d$). Note that $\omega_2 = 0$, with the components of ω_1 linearly independent over $\mathbb{Z}$, describes in the fully nonlinear case a resonance of higher multiplicity (hence the letter m).

– Perturbation: There are several specializations of the perturbative terms F (assumed to be analytic) which are of interest, if only to explore some issues while focusing on the main point, without dragging behind cumbersome and inessential difficulties. One well-known particular case (assuming full torsion for simplicity), is when the perturbation vanishes on the unperturbed invariant tori $p = q = 0, I = I_0$, labeled by I_0 (and parametrized by ϕ). This can be effected simply by considering (as was done in [A1]) perturbations of the form $F(q, \phi, I, p) = (\cos q - 1)f(q, \phi, I, p)$. Refinements are possible in this direction, and some will be considered in Chapter 3. At any rate, knowing the perturbed invariant tori, namely that they are just identical to the unperturbed ones, and that they exist for any frequency (not just Diophantine ones) does simplify some issues. Another direction in which the perturbation can be made special is by assuming partial or complete independence from the action variables. This is conceptually significant in the case of vanishing twist, where independence of the perturbation from some of the action variables has to be assumed in order for KAM theory to apply (see the appendix). Sometimes one assumes that F depends on the angles only ($F = F(q, \phi)$) simply for convenience; for instance this assumption does simplify explicit low-order computations (see Chapter 3). As a rule incorporating some p-dependence ($F = F(q, \phi, p)$), is not very significant conceptually, but leaving out I somehow is. This is so in the vanishing twist case as noted above, and also in the related mixed case. Consider indeed a mixed case with notation as in the last item, and assume that $F = F(q, \phi, I_1)$, *i.e.* it does not depend on the "linear" part I_2 of the action variables. Then the evolution of ϕ_2 is linear in the full system and we get a quasiperiodic perturbation of a fully nonlinear system. This is precisely the kind of systems which were investigated by B.V.Chirikov; indeed they arise rather naturally from a physical standpoint and lend themselves well to numerical experiments. This will be discussed again below (see §2.6 and the appendix).

Although other cases of interest and finer points could surely be added, we will here put an end to this terse review, which hopefully illustrates some of the interesting phenomena encapsulated in Hamiltonian (∗) and may give some feeling in the more general situations studied below.

Let us add a word about notation; we have tried to use the most convenient and often almost conventional one in each situation. This results in slight discrepancies along the text, but they should be harmless and obvious given the context. Note that we occasionally use the convenient French (or Bourbaki rather) pieces of notation]., . [, [., . [,] ., .] and [., .] for open and closed segments of a line. We finally mention that the labeling of a statement is given by that of the subsection in which it appears.

ACKNOWLEDGMENTS. This project developed over three years, and apart from the intrinsic interest of the subject, a strong impetus at the beginning came from the desire to understand the connections between some papers which appeared to very nearly contradict each other. The result of this effort should be apparent underneath, and is in part that of a collective process. In particular we are indebted to M.Rudnev and S.Wiggins for a one-year long correspondence (1996-97) which turned out to be as useful as it was enjoyable—at times comical. There were moments when we felt we were wading through a quagmire of vague irreconcilable assertions, and this resulted in reams of e-mails, some of which were as enlightening as they were contradictory. It is also a pleasure to thank P.Bernard, J.Cresson and L.Niederman for hours of useful conversations, as well several researchers in Barcelona, in particular A.Delshams and C.Simó, for sharing some of their insights with us. Finally, we are grateful to an anonymous referee for careful and critical reading of the original manuscript.

CHAPTER 1

Symplectic Geometry and the Splitting of Invariant Manifolds

In this first chapter we give an exposition of the *geometric* ideas involved in the study of the splitting of invariant manifolds. For the convenience of the reader, and also for the sake of fixing notation, the first two paragraphs shortly recall the necessary basic notions concerning symplectic geometry (§1.1) and hyperbolic sets in a general setting (§1.2). We refer to [AG] and [MDS] for detailed treatments of symplectic geometry, and to [HK] for an excellent general exposition of dynamical systems theory. We work everywhere with *Hamiltonian flows*, defined on a 2ℓ-dimensional symplectic manifold (M, Ω) by at least C^2 Hamiltonians; most applications will be to analytic systems. We leave it to the reader to convince himself that many properties can easily be adapted to other settings, in particular to symplectic maps instead of flows.

We introduce in Paragraphs 1.3–1.7 various notions of *angular splitting*, first at a linear level (splitting of two *linear* Lagrangian subspaces in a symplectic vector space), then for pairs of *Lagrangian submanifolds* of a symplectic manifold. We will distinguish between the *symplectic* notions, which can be defined without additional structure, and the *Euclidean* ones, which require in addition fixing an almost complex structure on the ambient manifold. The definitions are given first intrinsically, and then translated in local coordinates systems; we will stress the influence of the changes of coordinates on these local interpretations.

Paragraph 1.8 gives the definition and basic properties of hyperbolic tori in the Hamiltonian framework. We first introduce the *dynamical* definition of hyperbolicity, which does not require the knowledge of a particular normal form near the tori, and then take into account the symplectic geometry of the problem. The present study is thus general enough to be applied, including in a perturbative setting, to families of *non-KAM* hyperbolic tori, for instance the ones with Liouvillian rotation vectors (see [Y1]). Similar definitions are given in [Bol1], [Bol2], [BT].

Given a Hamiltonian H on M and an integer m satisfying $1 \leq m \leq \ell - 1$, a *m-hyperbolic* invariant torus $\mathcal{T}$ is an $(\ell - m)$-dimensional invariant torus in M such that the flow defined by H restricts to an ergodic flow on $\mathcal{T}$ (which in applications is generally conjugate to a transitive rotation) with exactly m strictly positive Lyapunov exponents, hence also m strictly negative ones (see the precise definition below). Let us mention that the case $m = 1$ is indeed special, as far as diffusion properties are concerned, and there are compelling reasons which lead to restrict certain aspects of the study to 1-hyperbolic tori. We have tried nevertheless to give general definitions and proofs for hyperbolic tori, most (but not all) of them being independent of the multiplicity m. It should be pointed out from the start that of course, if $m \neq \ell - 1$, a m-hyperbolic torus is *not* normally hyperbolic, for dimensional reasons, and that the case $m = \ell - 1$ corresponds to the well-known Birkhoff's (normally) hyperbolic periodic orbits. The theory of pseudo-hyperbolic manifolds (recalled in §1.2) makes it nevertheless possible in all cases to prove the existence of invariant stable and unstable manifolds for $\mathcal{T}$.

At the symplectic level, the torus $\mathcal{T}$ is assumed to be *isotropic*, and by the (dynamical) definition of hyperbolic objects, its invariant manifolds will be isotropic too and thus *Lagrangian*, since they are ℓ-dimensional. This Lagrangian character is probably the most important datum in our study, and will prove useful over and over again; in particular the various notions of homoclinic splitting for these invariant manifolds appear as particular cases of the preceding definitions (§§1.3–1.7). Again part of these results are obtained in [Bol1], [Bol2] and [BT] in a general context.

We describe in §1.9 a class of Hamiltonian systems which exhibit the main features that we want to analyze in the perturbative setting. That class generalizes both Hamiltonian (∗) and the normal forms one has to consider near (simple or multiple) resonances. We prove in particular some *symplectic straightening* theorems for the invariant manifolds, which in turn allow us in Chapter 2 to get simpler normal forms and to facilitate the application of KAM theory.

In Paragraph 1.10, we address the question of the existence of homoclinic orbits, in the perturbative setting, from the point of view of *Lagrangian intersection* theory. We generalize Eliasson's study [El] of the $m = 1$ case, and give very simple existence proofs for all multiplicities, based on the comparison of the cohomologies of the invariant manifolds. Our approach requires the knowledge of a homoclinic orbit biasymptotic to the fixed point in the averaged system, the existence of which is proved by usual variational methods, and we also assume (though it is not really necessary) the transversality of the intersection along this homoclinic. As far as the *existence* properties alone are concerned, variational methods can also be applied and lead to more general results (see [Bol1], [Bol2], [Fa]), whereas we believe that our method will be more efficient for the sake of *locating* the homoclinic orbits and proving the existence of genuine *hyperbolic* behaviour in the system.

Finally, Paragraph 1.11 is devoted to a geometric description of the splitting of the invariant manifolds of hyperbolic tori, mainly in the perturbative setting, for which we examine the connections between our definitions and the usual formalism (based essentially on the Poincaré-Melnikov integrals).

As a conclusion, we note that the notion of homoclinic splitting of manifolds has in fact several geometric and dynamical meanings. In particular it quantifies the *maximal distance* between two nearby hyperbolic tori connected by a *heteroclinic* orbit; passing from a homoclinic splitting to a heteroclinic distance is usually done by using suitable versions of the Implicit Function Theorem, which, due to the natural decomposition of the space in horizontal and vertical directions, requires only the purely *symplectic* interpretation. It also quantifies some *dynamical* parameters, such as the hyperbolicity of secondary invariants sets which appear, under additional assumptions on the torsion, in the neighborhood of transversal homoclinic tori (see [Cr1]), and this point of view necessitates the full geometric information and thus the *Euclidean* definition. Recent developments in symplectic topology make it possible to give more global definitions of the splitting, mainly based on Hofer's notion of *displacement energy*, but we will not enter into such considerations here. There are also interesting connections of this ubiquitous quantity with the notion of Peierls' barrier which is familiar in a variational setting ([Fa]).

1.1. Symplectic geometry: a short reminder

We let (M, Ω) denote a 2ℓ-dimensional smooth symplectic manifold and list some elementary notions of the attached geometry to fix notation.

1.1.1. First recall that a hypersurface $\mathcal{H} \subset M$ is always *coisotropic*, meaning that for all $x \in \mathcal{H}$, the symplectic orthogonal $(T_x\mathcal{H})^\perp$ is contained in $T_x\mathcal{H}$. One easily sees that $(T_x\mathcal{H})^\perp$ is one-dimensional and indeed is the kernel of the restriction of Ω_x to the tangent space $T_x\mathcal{H}$. Every hypersurface $\mathcal{H}$ is thus endowed with its *characteristic line-field* v, defined by $v(x) = (T_x\mathcal{H})^\perp$.

1.1.2. When $\mathcal{H}$ is a *regular* level set of a C^r function H ($r \geq 2$), this characteristic field is *orientable* and generated by the Hamiltonian vector field associated with H, defined *via* the duality $\iota_{X_H}\Omega = -dH$, where $\iota_{X_H}\Omega$ is the interior product of Ω by X_H. The manifold $\mathcal{H}$ is thus obviously invariant under the Hamiltonian flow. We will suppose in the sequel that $\mathcal{H} = H^{-1}(\{h\})$, where h is a regular value of H; we denote by X_H the vector field associated with H and by Φ the local flow of X_H. As usual Φ_t is the time-t diffeomorphism associated with Φ, defined on an open set of M, and we denote by ϕ the time-one map ($\phi = \Phi_1$).

1.1.3. It is well-known that Φ_t is *symplectic* on its domain: if $\Omega(s) = \Phi_s^*\Omega$ for $s \in [0, t]$, the derivative $\Omega'(s)$ is given by $\Omega'(s) = \Phi_s^*(L_{X_H}\Omega)$, and the Lie derivative $L_{X_H}\Omega = \iota_{X_H}d\Omega + d\iota_{X_H}\Omega = -dd\,H$ vanishes; thus $\Phi_t^*\Omega = \Omega$.

1.1.4. One frequently has to make use of *sections* for the Hamiltonian vector field X_H; these are simply codimension one submanifolds *of the ambient manifold* which are everywhere transverse to X_H. If Σ is a section of X_H contained in $\mathcal{H}$ (thus $(2\ell - 2)$-dimensional since $\mathcal{H}$ is $(2\ell - 1)$-dimensional) one easily sees that, denoting by $i_\Sigma : \Sigma \to M$ the canonical embedding, the form $\Omega_\Sigma = i_\Sigma^*\Omega$ induced by the symplectic form on Σ is nondegenerate. The pair (Σ, Ω_Σ) is thus a new symplectic manifold: this is an elementary example of a *symplectic reduction*.

1.1.5. Suppose that Σ_1 and Σ_2 are two sections of the flow contained in the same level $\mathcal{H}$, and connected by an orbit γ satisfying $x = \gamma(0) \in \Sigma_1$ and $y = \gamma(\tau) \in \Sigma_2$. By the transversality condition, there exist two neighborhoods O_1 and O_2 of x and y in the sections such that the flow Φ induces a *transition map* ψ from O_1 onto O_2. For any strictly positive smooth function μ on M, the new Hamiltonian $K = \mu(H - h)$ satisfies $\mathcal{H} = K^{-1}(\{0\})$ and $X_K(x) = \mu(x)\, X_H(x)$ for all $x \in \mathcal{H}$ (the function μ defines a reparametrization of the time). It is not difficult to find μ such that the transition map ψ is exactly given by the restriction to O_1 of the time-one map Ψ_1 of the flow associated with X_K, so one can always consider a transition map as the *time-one* map of a suitable Hamiltonian flow. One sees thus that such a map is *symplectic* if the sections are both endowed with their induced symplectic structure.

1.1.6. A symplectic manifold (M, Ω) is said to be *exact* when there exists a one-form λ such that $d\lambda = \Omega$; the form λ is then called a *symplectic potential* and is not unique.

Given two exact symplectic manifolds $(M_1, \Omega_1, \lambda_1)$ and $(M_2, \Omega_2, \lambda_2)$ an *exact* map from M_1 to M_2 is by definition a symplectic map g such that the closed one-form $g^*\lambda_2 - \lambda_1$ is exact, *i.e.* $g^*\lambda_2 - \lambda_1 = d\rho$, where ρ is a function on M_1. The composition of two exact maps is also exact.

1.1.7. Assuming that (M, Ω, λ) is exact, it is easy to check that the time-t map Φ_t is an exact map from its domain into M: note first that if $\lambda(s) = \Phi_s^*(\lambda)$,

$$\lambda'(s) = \Phi_s^*(L_{X_H}\lambda) = \Phi_s^*(\iota_{X_H} d\lambda + d(\iota_{X_H}\lambda))$$
$$= \Phi_s^*(d(-H + \iota_{X_H}\lambda)) = d\Phi_s^*(-H + \iota_{X_H}\lambda),$$

and thus $\Phi_t^*(\lambda) - \lambda = d\left(\int_0^t \Phi_s^*(-H + \iota_{X_H}\lambda)\, ds\right)$.

1.1.8. One sees also that if M is exact and if Σ is a section contained in $\mathcal{H}$, the form Ω_Σ is exact, being the exterior derivative of the induced form $i_\Sigma^*\lambda$: this gives the *induced* exact structure of Σ. Considering two sections Σ_1 and Σ_2 as in 1.1.5 above, endowed with their induced exact structures, one checks using the preceding argument that the transition map ψ between O_1 and O_2 is exact.

1.1.9. We will be especially concerned with the study of *Lagrangian submanifolds* of the given energy level $\mathcal{H}$. Recall first that a submanifold V of M is said to be *isotropic* when the form $i_V^*\Omega$ vanishes (where i_V is the inclusion map), or equivalently when for all $x \in V$ the tangent space T_xV is contained in its symplectic orthogonal $(T_xV)^\perp$. One sees that the dimension of an isotropic submanifold is at most ℓ. A submanifold V is *Lagrangian* when it is isotropic of maximal dimension, *i.e.* ℓ-dimensional; clearly then $T_xV = (T_xV)^\perp$ for all $x \in V$.

We will have to consider more generally *immersed* submanifolds of M, defined by injective immersions $f : \mathbf{V} \to M$, where $\mathbf{V}$ is a given manifold; this is indeed the case for the *global* invariant manifolds of a hyperbolic torus. We still denote them by $V = f(\mathbf{V})$ when there is no risk of confusion. One says that the immersion $f : \mathbf{V} \to M$ (or its image V) is isotropic when the pull-back $f^*\Omega$ vanishes, and Lagrangian when $\mathbf{V}$ is ℓ-dimensional and f isotropic. Of course a usual submanifold W of M is also an immersed submanifold, the map f being the canonical embedding i_W.

1.1.10. A fundamental property of Lagrangian submanifolds is their invariance under the Hamiltonian flow when they are contained in a level set of the Hamiltonian: if L is a Lagrangian submanifold contained in $\mathcal{H}$, $X_H(x)$ is in the tangent space T_xL for any $x \in L$. This is easy to see:

$$X_H(x) \in (T_xL)^\perp = T_xL.$$

One has a parallel result for immersed Lagrangian submanifolds.

1.1.11. Assuming again that (M, Ω, λ) is exact, an immersion $f : \mathbf{L} \to M$ (or its range $L = f(\mathbf{L})$) is said to be *exact* when it is *Lagangian* and when the *closed* form $f^*(\lambda)$ is exact, *i.e.* when $f^*\lambda = d\nu$, where ν is a function on $\mathbf{L}$. When L is an ℓ-dimensional submanifold of M, this simply means that $i_L^*\lambda = d\nu$, where ν is a function on L.

1.1.12. If f is an exact diffeomorphism of $(M_1, \Omega_1, \lambda_1)$ into $(M_2, \Omega_2, \lambda_2)$, and L is an exact submanifold of M_1, the direct image $f(L)$ is an exact submanifold of M_2. The analogous property holds true for immersed submanifolds.

1.1.13. Any cotangent bundle T^*V is an exact symplectic manifold: there exists a canonical one-form λ (the *Liouville* one-form) on T^*V characterized by the property that for every one-form α on V, $\alpha^*\lambda = \alpha$, where α is considered as a function from V into T^*V. It is easy to see that the derivative $\Omega = d\lambda$ is symplectic, and one usually endows T^*V with this symplectic form. In local cotangent coordinates (ξ_i, η_i) one finds that $\lambda = \sum_{i=1}^{\ell} \eta_i \, d\xi_i$ and thus $\Omega = \sum_{i=1}^{\ell} d\eta_i \wedge d\xi_i$. Using the characterization of λ one easily shows that if η is a given one-form on V, the submanifold $\mathrm{im}(\eta) \subset T^*V$ is Lagrangian iff η is closed, and exact iff η is exact.

1.1.14. Weinstein's symplectic tubular neighborhood theorem (see [MDS]) asserts that if L is a compact Lagrangian submanifold of the symplectic manifold (M, Ω) there exists a symplectic diffeomorphism between a neighborhood of L in M and a neighborhood of the null section of T^*L endowed with the Liouville structure described in 1.1.13 above. The neighborhoods of Lagrangian manifolds thus inherit all the properties of the cotangent bundles.

1.1.15. When a Hamiltonian H is given on the cotangent bundle T^*V, the *Hamilton-Jacobi* equation for a value h of H is the partial differential equation

$$\text{(HJ)} \qquad H \circ dS = h$$

with unknown function S. If S is a solution of (HJ), the image $\mathcal{L}_S = \mathrm{im}(dS)$ is an exact Lagrangian submanifold of M contained in the level set $\mathcal{H} = H^{-1}(\{h\})$, thus invariant under the Hamiltonian vector field. The *characteristic vector field* D_S for $\mathcal{L}_S$ is the image under the canonical projection $\pi : T^*V \to V$ of the restriction of X_H to $\mathcal{L}_S$, more precisely:

$$D_S(x) = T_{dS(x)}\pi\bigl(X_H(dS(x))\bigr).$$

The lifts to $\mathcal{L}_S$ of the orbits of D_S are thus the orbits of the field X_H contained in $\mathcal{L}_S$.

We stop here with this short and general reminder; more specific notions will be recalled and discussed in the rest of §1.

1.2. Hyperbolic invariant manifolds

We recall in this paragraph some standard and useful facts about pseudo-hyperbolic and normally hyperbolic invariant manifolds for general dynamical systems. Two excellent texts on these questions are [Y2] and [HPS]; more recently [W] provides a survey of the closely related work of Fenichel. Finally, we refer to [Ch] for very interesting new developments on the subject. We will essentially be interested in the *dynamical* consequences of the existence of invariant manifolds, and we therefore limit ourselves to the usual statements in the C^r category, $1 \leq r \leq \infty$. Part of the results are still valid in the analytic category, namely those concerning analytic hyperbolic tori and their invariant manifolds, but not all of them: even in analytic systems the normally hyperbolic invariant manifolds are generally not analytic but only C^r, the order r depending on the ratio between the longitudinal and transversal contraction and dilatation rates along the manifold. Note that C^2 would be enough for most applications of the theory to diffusion problems.

1.2.1. We first consider a C^r $(r \geq 1)$ diffeomorphism f of a C^∞ manifold M which leaves a compact C^r submanifold V invariant. Given a Riemannian metric $\| \ \|$ on M, and a subbundle E of $T_V M$ invariant under Tf, we denote by $B_a(E)$ the unit ball in the fiber E_a, for $a \in V$, and write:

$$M(Tf_{|E}) = \mathrm{Sup}\,\bigl\{\|T_a f_{|E}\| \mid a \in V\bigr\}, \quad m(Tf_{|E}) = \bigl(M(Tf_{|E}^{-1})\bigr)^{-1},$$

where, as usual:
$$\|T_a f_{|E}\| = \mathrm{Sup}\{\|T_a f(u)\| \mid u \in B_a(E)\}.$$

DEFINITION. Given $\rho > 0$, the manifold V is said to be ρ-pseudo-hyperbolic for f if there exist two continuous subbundles E^+, E^- of the tangent bundle $T_V M$ and a Riemannian metric $\|\ \|$ on M which satisfy the conditions:
i) $T_V M = E^+ \oplus E^-$;
ii) $Tf(E^+) \subset E^+$, $Tf(E^-) \subset E^-$;
iii) $M(Tf_{|E^+}) < \rho$, $m(Tf_{|E^-}) > \rho$.

The metric $\|\ \|$ is then said to be ρ-adapted to f and V. For a given integer $k \geq 1$, let us denote by $E^r(D^k, M)$ the space of C^r embeddings of the open unit ball $D^k \subset \mathbb{R}^k$ into M. The fundamental result for pseudo-hyperbolic invariant manifolds is as follows:

THEOREM (The local stable manifold theorem). *Suppose that V is ρ-pseudo-hyperbolic for f, and let $\|\ \|$ be any Riemannian metric on M with associated distance d. Let k be the dimension of E^+. Then there exists a continous function $\mathcal{P}$ from V to $E^r(D^k, M)$ such that, if Δ_a is the embedded ball $\mathcal{P}(a)(D^k)$:*
i) $a \in \Delta_a$ for all $a \in V$;
ii) the ball Δ_a is tangent to the fiber E_a^+ at the point $a \in V$;
iii) there exists $C > 0$ such that, for all $a \in V$, $b \in \Delta_a$, $n \in \mathbb{Z}_+$: $d(f^n(a), f^n(b)) \leq C\rho^n$;
iv) if $a \neq a'$ are two points in V, the intersection $\Delta_a \cap \Delta_{a'}$ is relatively open in both Δ_a and $\Delta_{a'}$.

We refer to [Y2] for a short proof. Note that *iii)* can be rephrased as:
iii') for all $\rho \in \,]M(Tf_{|E^+}), m(Tf_{|E^-})[$, $a \in V$ and $b \in \Delta_a$: $d(f^n(a), f^n(b)) = o(\rho^n)$.

If one adds the assumption that $\rho < 1$, the ball Δ_a can be dynamically interpreted as a *local ρ-stable manifold for a*: the iterates of each point in Δ_a are positively asymptotic to the ones of a, with an $o(\rho^n)$ convergence rate. Although this assumption is not necessary, it will be made in the sequel for the sake of simplicity. From now on we will call Δ_a the ρ-stable leaf (or stable leaf for short) associated with a, and denote it by Δ_a^+ so as to underline its dynamical interpretation.

1.2.2. Even in this weak form, the theorem provides not only the existence of these local invariant manifolds, but also gives preliminary indications on their mutual arrangements (see *iv)*). Nevertheless it is too weak for our purposes: one would like to prove the existence of a stable manifold attached to V itself, not only to each of its points. A well-known case of existence is that of *normally hyperbolic manifolds* (see below), but the invariant tori that we have to consider generally do *not* enter into that category, for dimensional reasons and due to the symplectic character of the flow. We thus introduce a new definition, to which we will refer when regularity properties of the invariant manifolds are needed. With the same notation as above, remark first that if V is ρ-pseudo-hyperbolic for f, and if one furthermore assumes $TV \subset E^-$ to ensure a good asymptotic behaviour of the points on V, then the stable leaves Δ_a^+ are pairwise disjoint. This comes from Theorem 1.2.1 together with the fact that, given $c > 0$, the distance between the n-th iterates of two points in V cannot be less than $2c\rho^n$ for all n. One can thus consider the

union
$$W^+(V) = \bigcup_{a \in V} \Delta_a^+,$$
which is obviously the ρ-stable manifold of V, and ask for regularity properties of $W^+(V)$.

DEFINITION. Let $\rho \in]0,1[$ and consider an integer $l \geq 1$. The invariant manifold V is said to be (ρ, l) regularly hyperbolic for f if there exist two continuous subbundles E^+, E^- of the tangent bundle $T_V M$, and a Riemannian metric $\|\ \|$ on M which satisfy the conditions:
i) $T_V M = E^+ \oplus E^-$;
ii) $Tf(E^+) \subset E^+$, $Tf(E^-) \subset E^-$;
iii) $TV \subset E^-$;
iv) $M(Tf_{|E^+}) < \rho$, $m(Tf_{|E^-}) > \rho$;
v) $W^+(V)$ is a C^l-submanifold of M.

Note finally that the existence of *unstable* invariant manifolds can be deduced from the preceding theorems applied to f^{-1}, when of course the corresponding assumptions are met. In that case, we will denote by Δ_a^- the stable leaf associated with a and relative to f^{-1}, and call it the *unstable leaf* associated with a. In the regular case, the union of the unstable leaves is the (local) unstable manifold for V and is denoted by $W^-(V)$.

1.2.3. We now recall the definition of normal hyperbolicity.

DEFINITION. Let s be an integer, $1 \leq s \leq r$. The manifold V is said to be s-normally-hyperbolic for f if there exist two continuous subbundles E^+, E^- of $T_V M$ and a Riemannian metric $\|\ \|$ on M satisfying the following conditions:
i) $T_V M = E^+ \oplus TV \oplus E^-$;
ii) $Tf(E^+) \subset E^+$, $Tf(E^-) \subset E^-$;
iii) $M(Tf_{|E^+}) < m(Tf_{|TV})^k \leq 1 \leq M(Tf_{|TV})^k < m(Tf_{|E^-})$, for $k = 1, \ldots, s$.

The (un)stable manifold theorem in this case takes the following refined form (see [HPS]):

THEOREM (The local invariant manifolds theorem). *Suppose V is s-normally-hyperbolic for f. Let*
$$M_+ = M(Tf_{|E^+}), \quad m_T = m(Tf_{|TV}), \quad M_T = M(Tf_{|TV}), \quad m_- = m(Tf_{|E^-}).$$
Then, if d is the distance associated with a given Riemannian metric on M, there exists a neighborhood N of V such that the following properties hold:
i) *the sets*
$$W^+(V) = \left\{ b \in N \mid \forall \rho \in]M_+, m_T[, \lim_{n \to \infty} \rho^{-n} d(f^n(b), V) = 0 \right\}$$
$$W^-(V) = \left\{ b \in N \mid \forall \rho \in]M_T, m_-[, \lim_{n \to \infty} \rho^{-n} d(f^{-n}(b), V) = 0 \right\}$$
are C^s submanifolds of N;
ii) *at any point $a \in V$, the manifold $W^+(V)$ is tangent to $E_a^+ \oplus TV$ and $W^-(V)$ is tangent to $E_a^- \oplus TV$;*
iii) *$W^+(V)$ is the maximal set contained in N satisfying $f(W^+(V)) \subset W^+(V)$ and $W^-(V)$ is the maximal set contained in N satisfying $f^{-1}(W^-(V)) \subset W^-(V)$;*

iv) there exist two f-invariant C^s foliations of $W^+(V)$ and $W^-(V)$, the leaves of which are stable and unstable leaves Δ_a^+, Δ_a^- associated with the points of V, these leaves are tangent to the fibers E_a^+ and E_a^- at each point $a \in V$;

v) if g is close enough to f in the C^s topology, there exist a C^s normally invariant manifold V' which is C^s-close to V; the stable and unstable manifolds $W^+(V')$ and $W^-(V')$ of V' are C^s-close to that of V, and the stable and unstable leaves of the points of V' are C^s-close to the ones associated with the points of V.

As usual, one deduces the existence of the *global* stable or unstable manifolds by taking the inverse or direct images of the local ones. In the case of flows on M, one defines the continuous version of the above notions and easily gets the analogous results *e.g.* by using the associated time-one maps.

1.3. Angles of Lagrangian planes: the symplectic viewpoint

Here we set off on our way to study various forms of splitting of Lagrangian manifolds. This paragraph and the next one are concerned with the linear situation, *i.e.* with Lagrangian planes in a symplectic vector space. The globalization of the notions to Lagrangian submanifolds of a symplectic manifold is discussed in Paragraph 1.6. It is useful to distinguish between purely symplectic notions and those which in addition assume the existence of a Euclidean structure. In this paragraph, we introduce the basic purely symplectic "angular" notions for pairs of Lagrangian subspaces of a symplectic vector space. In the next paragraph, we will use an additional compatible Hermitian structure in order to define *Euclidean* angles. Paragraph 1.5 gives the basis of the attending "calculus" in matrix terms, with a particular attention to the variance of the objects with respect to symplectic transformations. This will be crucial in particular when we specialize the various notions to the perturbative setting of Chapter 2.

1.3.1. Let (W, Ω) be a symplectic vector space of dimension 2ℓ. It is well-known that one can identify W with $\mathbb{R}^{2\ell}$, with coordinates (u_i, v_j), in such a way that the symplectic form Ω reads $\Omega = \sum_{i=1}^{\ell} dv_i \wedge du_i$. We will explicitly make this identification below and denote by H_0 (horizontal) and V_0 (vertical) the Lagrangian plane $\mathbb{R}^\ell \times \{0\}$ and $\{0\} \times \mathbb{R}^\ell$ respectively.

If H and V are two transverse Lagrangian subspaces of $\mathbb{R}^{2\ell}$, the dual space H^* of H can be identified with V by means of the symplectic form: for $v \in V$, simply define $v^*(u) = \Omega(u, v)$ for $u \in H$, and use the fact that V is Lagrangian to prove that this gives rise to an isomorphism.

The set of all Lagrangian subspaces L of $\mathbb{R}^{2\ell}$ will be denoted by $\Lambda(\mathbb{R}^{2\ell}, \Omega)$, or simply Λ. The symplectic group $Sp(2\ell, \mathbb{R})$ acts transitively on Λ, and the stabilizer of H_0 is the subgroup $C \subset Sp(2\ell, \mathbb{R})$ of all symplectic lifts of linear isomorphisms of H_0. So $C \simeq Gl(\ell, \mathbb{R})$ and the Lagrangian-Grassmannian Λ is thus an algebraic variety of dimension $\ell(\ell+1)/2$ which is isomorphic to the quotient $Sp(2\ell, \mathbb{R})/Gl(\ell, \mathbb{R})$; we will see another identification below.

The construction of local coordinates on the manifold Λ is classical. Forgetting first about the Lagrangian character of the spaces, consider a pair of ℓ-dimensional transverse spaces H and V of $\mathbb{R}^{2\ell}$ and denote by p_H and p_V the projections on H and V associated with the direct sum $\mathbb{R}^{2\ell} = H \oplus V$. Let then L be an ℓ-dimensional subspace transverse to V. The restriction to L of the projection p_H is an isomorphism of L onto H, by transversality of L and V; denote its inverse by $q : \text{H} \to \text{L}$. Then L is the *graph* of the linear map $j_\text{L} = p_\text{V} \circ q$ from H to V (identifying $\text{H} \oplus \text{V}$ with $\text{H} \times \text{V}$); we say that the linear map j_L is the *equation of* L *relative to* H *and* V. Since $\text{L} = \{a + j_\text{L}(a) \mid a \in \text{H}\}$, where a and $j_\text{L}(a)$ are considered as elements of $\mathbb{R}^{2\ell}$, one may also consider the map $l_\text{L} : \text{H} \to \mathbb{R}^{2\ell}$ defined by $l_\text{L}(a) = a + j_\text{L}(a)$, giving a linear embedding of H onto L which we call the *natural embedding* for L, relative to H and V.

Assume now that H and V are *Lagrangian* subspaces of $\mathbb{R}^{2\ell}$. Given an ℓ-dimensional subspace L (not necessarily Lagrangian), one can now define a bilinear form on H through the relation

$$\gamma_\text{L}(a,b) = \Omega(a, j_\text{L}(b)), \qquad a \in \text{H}, \ b \in \text{H};$$

the form γ_L is the *cooordinate* of L relative to H and V. Note that using the identification of H^* with V, this is the classical identification of a linear map from H to H^* to a bilinear form on H; so the correspondence $\text{L} \mapsto \gamma_\text{L}$ is one-to-one.

The crucial property is that γ_L *is symmetric if and only if* L *is Lagrangian*. Indeed note that

$$\Omega(a + j_\text{L}(a), b + j_\text{L}(b)) = \Omega(a, j_\text{L}(b)) + \Omega(j_\text{L}(a), b) = \gamma_\text{L}(a,b) - \gamma_\text{L}(b,a)$$

since H and V are Lagrangian, and the symmetry of γ_L is thus equivalent to the vanishing of Ω on L.

Let us denote by Λ_V the set of all Lagrangian spaces which are transverse to the space V; clearly Λ_V is an open subset of Λ (classically refered to as a "Schubert cell"). The correspondence $\text{L} \mapsto \gamma_\text{L}$ allows us to identify Λ_V with the space $\mathcal{S}(\text{H})$ of *symmetric bilinear forms* on H. Since $\mathcal{S}(\text{H})$ is isomorphic to $\mathbb{R}^{\ell(\ell+1)/2}$, this identification can be considered as a coordinate map over the open set Λ_V. Varying V and H, we obtain in this way an atlas for the Lagrangian-Grassmannian Λ.

The following remark will be useful in the sequel. We have seen that V is naturally identified with the dual space H^*; on the other hand one can consider the quadratic form Q_L on H associated with γ_L (*i.e.* $Q_\text{L}(a) = \tfrac{1}{2}\gamma_\text{L}(a,a)$), and its derivative dQ_L from H to $\text{H}^* = \text{V}$. Then one finds that

$$dQ_\text{L}(a)(b) = \gamma_\text{L}(a,b) = \gamma_\text{L}(b,a) = \Omega(b, j_\text{L}(a)),$$

so that $dQ_\text{L}(a)$ is identified with $j_\text{L}(a)$. In other words the Lagrangian space L is also the graph of the derivative of the quadratic form Q_L. The form Q_L is said to be a *generating function* for L.

Finally one can go to coordinates and introduce matrices for the preceding objects. Given a basis $B_\text{H} = (e_1, \ldots, e_\ell)$ of H it is possible to find a basis $B_\text{V} = (f_1, \ldots, f_\ell)$ of V such that the union $B = (e_1, \ldots, e_\ell, f_1, \ldots, f_\ell)$ is symplectic: this amounts to the condition $\Omega(e_i, f_j) = \delta_{ij}$, since H and V are Lagrangian. One sees easily that the matrices of γ_L relative to B_H and of j_L relative to B_H and B_V coincide. Writing $\mathbf{J}_\text{L}$ for that common matrix, the matrix of l_L relative to B_H and B is thus $\begin{bmatrix} \mathbf{I} \\ \mathbf{J}_\text{L} \end{bmatrix}$, where $\mathbf{I}$ is the $(\ell \times \ell)$ identity matrix.

1.3.2. The symplectic angular notions. The first remark concerning the possibility of defining symplectic *angles* is disappointing: the action of the symplectic group on Λ^2, which would be the most natural thing to consider in order to define symplectic angles, does not lead to any interesting notion. Indeed, given two pairs (L_1, L_2) and (L'_1, L'_2) of Lagrangian spaces, a necessary and sufficient condition for the existence of a symplectic isomorphism ψ of $\mathbb{R}^{2\ell}$ such that $\psi(L_1) = L'_1$ and $\psi(L_2) = L'_2$ is simply that the intersections $L_1 \cap L_2$ and $L'_1 \cap L'_2$ have the same dimension. In this respect this is not different from the action of the full linear group $Gl(2\ell, \mathbb{R})$: the *symplectic angle* of two Lagrangian spaces is just the dimension of their intersection. One thus has to refine the primitive notion, which we do by considering *linear Lagrangian embeddings* of an ℓ-dimensional vector space E into $\mathbb{R}^{2\ell}$, instead of abstract Lagrangian subspaces. This leads to the

DEFINITION. Consider two linear Lagrangian embeddings l_1 and l_2 of two ℓ-dimensional vector spaces E_1, E_2 into $\mathbb{R}^{2\ell}$. The symplectic angular form of the pair (l_1, l_2) is the bilinear form $\mathcal{A}_s$ on $E_1 \times E_2$, defined by

$$\mathcal{A}_s(a, b) = \Omega(l_1(a), l_2(b)), \qquad a \in E_1, \ b \in E_2.$$

The geometric significance of the angular form $\mathcal{A}_s$ is quite weak and does not go much beyond the following transversality result, whose proof is straightforward.

PROPOSITION. *Let $L_i = l_i(E_i)$, for $i = 1, 2$. Then, denoting by $\mathrm{Ker}_1 \mathcal{A}_s$ the set of $a \in E_1$ such that $\mathcal{A}_s(a, b) = 0$ for all $b \in E_2$, one has:*

$$\mathrm{Ker}_1 \mathcal{A}_s = l_1^{-1}(L_1 \cap L_2),$$

with an analogue for $\mathrm{Ker}_2 \mathcal{A}_s$. The spaces L_1 and L_2 are thus transverse in $\mathbb{R}^{2\ell}$ if and only if the symplectic angular form is nondegenerate.

The form $\mathcal{A}_s$ cannot be defined for a pair of Lagrangian subspaces (working with embeddings amounts to fixing bases for the two spaces), but one encounters frequently situations where such embeddings are "natural". The most important one is given by the previous construction of coordinates on Λ, when a decomposition $\mathbb{R}^{2\ell} = H \oplus V$ is given from the start. The correspondence $L \mapsto l_L$ identifies Λ_V with the set of linear Lagrangian embeddings of H into $\mathbb{R}^{2\ell}$ whose ranges are transverse to V. As a consequence, if a pair (L_1, L_2) in Λ_V^2 is given, one can define as above the symplectic angular form of the associated pair of embeddings (l_{L_1}, l_{L_2}).

DEFINITION. Let (L_1, L_2) be a pair of elements of Λ_V. The *splitting form* $\sigma_{HV}(L_1, L_2)$ of the pair (L_1, L_2) relative to the subspaces H and V is by definition the symplectic angular form $\mathcal{A}_s$ of the two natural embeddings l_{L_1} and l_{L_2} relative to H and V (here of course $E_1 = E_2 = H$). Given a basis B_H of H, the matrix $\mathbf{S}$ in B_H of the bilinear form $\sigma_{HV}(L_1, L_2)$ is the *splitting matrix* of the pair (L_1, L_2), relative to H, V and B_H.

As one would expect, the splitting form is directly related to the coordinates relative to H and V:

PROPOSITION. *Let (L_1, L_2) be a pair of elements of Λ_V. The splitting form of the pair (L_1, L_2) is given by the difference of the coordinates of L_1 and L_2 relative to H and V:*

$$\sigma_{HV}(L_1, L_2) = \gamma_{L_2} - \gamma_{L_1}.$$

It is thus a symmetric form on H.

PROOF. It amounts to an easy check. Writing σ for $\sigma_{HV}(L_1, L_2)$, we get, for $a \in H$ and $b \in H$:

$$\begin{aligned}\sigma(a,b) &= \Omega(l_{L_1}(a), l_{L_2}(b)) = \Omega(a + j_{L_1}(a), b + j_{L_2}(b)) \\ &= \Omega(a, j_{L_2}(b)) + \Omega(j_{L_1}(a), b) = \Omega(a, j_{L_2}(b)) - \Omega(a, j_{L_1}(b)) \\ &= \gamma_{L_2}(a,b) - \gamma_{L_1}(a,b),\end{aligned}$$

where the Lagrangian character of L_1 and L_2 was used in the third equality. $\square$

One can also consider the quadratic form on H associated with the splitting form $\sigma_{HV}(L_1, L_2)$ and the preceding proposition shows that it is also the difference $Q_{L_2} - Q_{L_1}$ of the generating functions of L_1 and L_2 relative to H and V. The splitting matrix in a basis B is thus nothing more than (twice) the difference of the matrices of these two quadratic forms, expressed in B. Note that a splitting matrix is always *symmetric*.

One would then like to interpret the splitting form in a more quantitative way. As far as transversality properties alone are concerned, one finds as before that L_1 and L_2 are transverse in $\mathbb{R}^{2\ell}$ if and only if the splitting form is nondegenerate, with $\dim(L_1 \cap L_2) = \dim \mathrm{Ker}(\sigma_{HV}(L_1, L_2))$. But even in $\mathbb{R}^2$, where each line is a Lagrangian subspace, there is *no* relation between the splitting form of two lines and (for instance) their usual angle. An element $L \in \Lambda_{V_0}$ is given by an equation $j_L(a) = \gamma a$, with $\gamma \in \mathbb{R}$ and $a \in H_0 = \mathbb{R}$ (so its usual equation reads $y = \gamma x$), and the form γ_L associated with L is simply given by its matrix $[\gamma]$ relative to the basis $e_1 = 1$ of $\mathbb{R}$. The splitting matrix of a pair (L_1, L_2) of $\Lambda_{V_0}^2$, with equations $j_{L_k}(a) = \gamma_k a$ $(k = 1, 2)$, is thus just the difference $[\gamma_2 - \gamma_1]$. The angle is $\theta = \tan^{-1}(\gamma_2) - \tan^{-1}(\gamma_1) \in \,]-\pi/2, \pi/2]$ (modulo π) and it is for instance possible to let θ go to infinity while keeping the matrix $[\gamma_2 - \gamma_1]$ constant.

By contrast, the knowledge of *both* coordinates γ_{L_1} and γ_{L_2} completely determines the relative positions of the two subspaces. The best way to see this will be to introduce first the notion of Euclidean angles for Lagrangian subspaces, and then to show how to compute their determinations from these data.

1.4. Angles of Lagrangian planes: the Euclidean viewpoint

We now assume that a compatible *complex structure* is given on the ambient symplectic vector space (W, Ω), meaning that W is identified with $\mathbb{C}^\ell$, with coordinates $w_j = u_j + i\, v_j$. The Hermitian form Ξ is given by

$$\Xi(w, w') = \sum_{j=1}^{\ell} w_j \overline{w}'_j = \Pi(w, w') + i\,\Omega(w, w')$$

for w and w' in $\mathbb{C}^\ell$, where Π is the usual scalar product of $\mathbb{R}^{2\ell}$. The identification $W \simeq \mathbb{R}^{2\ell} \simeq \mathbb{C}^\ell$ is explicitly used below.

The three groups relative to the forms Ξ, Π and Ω are of interest in the study: they are respectively the unitary group $U(\ell, \mathbb{C})$, the Euclidean orthogonal group $O(2\ell, \mathbb{R})$ and the symplectic group $Sp(2\ell, \mathbb{R})$. We recall the following useful relations:

$$U(\ell, \mathbb{C}) = O(2\ell, \mathbb{R}) \cap Sp(2\ell, \mathbb{R}) = Gl(\ell, \mathbb{C}) \cap O(2\ell, \mathbb{R}) = Gl(\ell, \mathbb{C}) \cap Sp(2\ell, \mathbb{R}).$$

Using the expression of Ξ, one sees that an ℓ-dimensional subspace L is Lagrangian if and only if every $\mathbb{R}$-basis of L, orthonormal for Π, is a *unitary* $\mathbb{C}$-basis of $\mathbb{C}^\ell$ for the hermitian complex structure. In this case, the union $B \cup iB$ is a symplectic Π-orthonormal basis of $\mathbb{R}^{2\ell}$. This is also equivalent to the equality $i\,\mathrm{L} = \mathrm{Orth}_\Pi(\mathrm{L})$, where $\mathrm{Orth}_\Pi(\mathrm{L})$ is the ℓ-dimensional subspace Π-orthogonal to L. As a consequence, the unitary group $U(\ell, \mathbb{C})$ acts transitively on Λ, and the stabilizer of the subspace H_0 is $O(\ell, \mathbb{R})$. The manifold Λ is thus also isomorphic to the quotient $U(\ell, \mathbb{C})/O(\ell, \mathbb{R})$.

1.4.1. Let $\mathbf{G}_\ell$ denote the Grassmannian of all ℓ-dimensional subspaces of $\mathbb{R}^{2\ell}$. Given a pair $(\mathrm{L}_1, \mathrm{L}_2)$ in $\mathbf{G}_\ell^2$, one defines as usual the Euclidean angle of $(\mathrm{L}_1, \mathrm{L}_2)$ as the orbit of $(\mathrm{L}_1, \mathrm{L}_2)$ under the diagonal action of the orthogonal group $O(2\ell, \mathbb{R})$ on $\mathbf{G}_\ell^2$, so the space of Euclidean angles is the quotient space $\mathbf{G}_\ell^2/O(2\ell, \mathbb{R})$. This notion is valid in particular when $(\mathrm{L}_1, \mathrm{L}_2)$ is a pair of *Lagrangian* spaces, but one easily sees that in that case the intersection of the orbit of the pair with Λ^2 is exactly the orbit of the pair under the diagonal action of the *unitary* group $U(\ell, \mathbb{C})$ on Λ^2. As we are only concerned with the study of Lagrangian objects, the following definition is natural.

DEFINITION. *Let $(\mathrm{L}_1, \mathrm{L}_2)$ be a pair of Lagrangian subspaces of $\mathbb{C}^\ell$. The Euclidean angle $\mathbf{A}(\mathrm{L}_1, \mathrm{L}_2)$ of $(\mathrm{L}_1, \mathrm{L}_2)$ is the orbit of the pair under the diagonal action of $U(\ell, \mathbb{C})$ on Λ^2. The space of Euclidean angles of Lagrangian spaces in $\mathbb{R}^{2\ell}$ is thus the quotient $\Lambda^2/U(\ell, \mathbb{C})$.*

Let us now be more explicit. It is useful to note first that given a Lagrangian space L, the equation j_L relative to H_0 and V_0 extends uniquely to $\mathbb{C}^\ell$ as a *complex* linear map, *i.e.* there exists a unique complex endomorphism f of $\mathbb{C}^\ell$ such that $f(a) = j_\mathrm{L}(a) \in \mathrm{V}_0$, for all $a \in \mathrm{H}_0$. One easily sees that f has the form $i\,J_\mathrm{L}$, where J_L is a *self-adjoint* endomorphism of $\mathbb{C}^\ell$ sending H_0 into H_0. This simply means that the matrix of J_L in a unitary basis B_{H_0} of $\mathbb{C}^\ell$ contained in H_0 is real and symmetric: indeed this is just the matrix of γ_L relative to B_{H_0}. The map $I + i\,J_\mathrm{L}$ is thus clearly the unique extension to $\mathbb{C}^\ell$ of the embedding l_L as a complex isomorphism, sending H_0 onto L. We may now state the central result of this paragraph.

PROPOSITION. *Let L be in Λ. There exists a unique unitary isomorphism $\varphi \in U(\ell, \mathbb{C})$, diagonalizable relatively to a unitary basis contained in H_0, with eigenvalues of the form $\rho = e^{i\theta}$, $\theta \in \,]-\pi/2, \pi/2]$, such that $\varphi(\mathrm{H}_0) = \mathrm{L}$.*

If L is transverse to V_0, and if γ is the coordinate of L relative to H_0 and V_0, φ and γ are diagonalizable in the same unitary basis contained in H_0, and if α is an eigenvalue of γ, the corresponding eigenvalue for φ is given by $\rho = (1 + i\,\alpha)/|1 + i\,\alpha|$.

PROOF. Suppose first that L is transverse to V_0. We consider the self-adjoint endomorphism J fixing H_0 such that $I + i\,J$ sends H_0 onto L. Let $B_{\mathrm{H}_0} = (a_1, \ldots, a_\ell)$ be a unitary basis of eigenvectors of J, contained in H_0, and let α_k be the (real) eigenvalue corresponding to a_k. Then the a_k are also eigenvectors for $I + i\,J$ and $I - i\,J$, with eigenvalues $1 + i\,\alpha_k$ and $1 - i\,\alpha_k$ respectively. The product $P = I + J^2 = (I + i\,J) \circ (I - i\,J)$ is thus symmetric and positive definite, with eigenvalues $1 + \alpha_k^2$. Consider the square root R of P defined by the diagonal matrix $\mathrm{diag}(\sqrt{1 + \alpha_1^2}, \ldots, \sqrt{1 + \alpha_l^2})$ in the basis B_{H_0}, and define $\varphi = (I + i\,J) \circ R^{-1}$. Since R maps H_0 onto itself, φ sends H_0 onto L, and φ is clearly unitary, with eigenvalues $(1 + i\alpha_k)/|1 + i\alpha_k|$ of the required form (i.e. with $\theta_k \in]-\pi/2, \pi/2[$). This proves the existence of φ when L is transverse to V_0. Since the matrix of γ in B_{H_0} is the same as the one of J, the second assertion is also proved.

If L is not transverse to V_0 let $E = i(V_0 \cap L) \subset H_0$. Since L is Lagrangian, E is Π-orthogonal to the projection $F = p_{H_0}(L)$. One can thus decompose the space $\mathbb{C}^\ell$ as the sum $G_\| \oplus G_\perp$, with $G_\| = E \oplus iE$ and $G_\perp = F \oplus iF$. The spaces $G_\|$ and $G_\perp$ are orthogonal complex subspaces of $\mathbb{C}^\ell$, and the intersection $l = L \cap G_\perp$ is Lagrangian (for the induced structure) and transverse to iF. Using the previous case, we find a unitary isomorphism $\varphi_\perp$ of $G_\perp$ sending F onto l, and diagonalizable in a unitary basis of F with eigenvalues in the required form. Let $\varphi_\|$ be the restriction to $G_\|$ of the product $u \mapsto iu$. One sees easily that the unitary morphism $\varphi = \varphi_\perp \oplus \varphi_\|$ has the required properties, the number of θ_k equal to $\pi/2$ being the dimension of $V_0 \cap L$.

As for uniqueness, assume φ is a unitary map as in the proposition. Then consider a unitary basis $B_{H_0} = (a_1, \ldots, a_\ell)$ of eigenvectors of φ, contained in H_0, and let ρ_k be the (complex) eigenvalue corresponding to a_k. One sees that $\rho_k a_k$ is in the intersection of the complex line $\mathbb{C} a_k$ with L_2, which is thus at least one-dimensional, and exactly one-dimensional since L is Lagrangian. This condition determines ρ_k in the required form and it is then easy to obtain the uniqueness of φ. See [Nic] for further information. □

1.4.2. The following consequence of the above will be relevant for our purpose:

COROLLARY. *Let (L_1, L_2) be a pair of Lagrangian subspaces of $\mathbb{R}^{2\ell}$. There exists a unique unitary isomorphism $\varphi_{L_1 L_2} \in U(\ell, \mathbb{C})$, diagonalizable relatively to a unitary basis contained in L_1, with eigenvalues of the form $\rho = e^{i\theta}$, $\theta \in]-\pi/2, \pi/2]$, such that $\varphi_{L_1 L_2}(L_1) = L_2$. If ϕ is a unitary isomorphism with $L'_i = \phi(L_i)$ for $i = 1, 2$, then $\varphi_{L'_1 L'_2} = \phi \circ \varphi_{L_1 L_2} \circ \phi^{-1}$.*

PROOF. If φ_0 is a unitary isomorphism mapping H_0 onto L_1, the set $U(L_1, L_2)$ of unitary isomorphisms sending L_1 onto L_2 is given by

$$U(L_1, L_2) = \{\varphi_0 \circ \varphi \circ \varphi_0^{-1} \mid \varphi \in U(H_0, \varphi_0^{-1}(L_2))\}$$

where $U(H_0, \varphi_0^{-1}(L_2))$ is the set of unitary isomorphisms mapping H_0 onto $\varphi_0^{-1}(L_2)$. It is then clear how to deduce the corollary from the proposition. □

One can thus identify the Euclidean angle defined by the pair (L_1, L_2) with the conjugacy class in $U(n, \mathbb{C})$ of the unitary morphism $\varphi_{L_1 L_2}$. This means that the space of Euclidean angles of Lagrangian spaces can in turn be identified with the space of conjugacy classes of unitary morphisms with eigenvalues as in the proposition above (here we use that a unitary basis $\mathbb{R}$-spans a Lagrangian space in $\mathbb{R}^{2\ell}$). We are thus led to the

DEFINITION. *Let $(L_1, L_2) \in \Lambda^2$ be a pair of Lagrangian planes. We say that the unitary morphism $\varphi_{L_1 L_2}$ mapping L_1 onto L_2 is a* representative *of the Euclidean angle* **A** *of the pair (L_1, L_2). The spectrum $S = (\rho_1, \ldots, \rho_k)$ (ordered counterclockwise in the right half of the unit circle) of $\varphi_{L_1 L_2}$ is the* measure *of* **A** *and the pair $((\rho_1, \ldots, \rho_k), (E_1, \ldots, E_k))$ formed by the measure and the collection of the associated eigenspaces we call a* determination *of* **A**.

Note that due to the fact that all representatives are conjugate, the measure of an angle is uniquely defined, whereas the determination depends on the choice of the pair of spaces, since the eigenspaces $E_1, \ldots, E_k$ are contained in the first member of the pair.

1.5. Symplectic isomorphisms, angles and splitting forms

We collect in this paragraph some explicit matrix computations and especially examine the effect of symplectic isomorphisms on the preceding angular notions. Such situations will arise naturally in the sequel in connection with (normalizing) canonical transformations induced for instance by time-one maps of Hamiltonian flows (Lie method).

1.5.1. The variance formula for the coordinates.
We return to explicit coordinates on the open cell Λ_V of the Lagrangian-Grassmannian Λ. That is we use the "natural" decomposition $\mathbb{R}^{2\ell} = H \oplus V$ and fix a symplectic basis $B = (e_1, \ldots, e_\ell, f_1, \ldots, f_\ell)$ of $\mathbb{R}^{2\ell}$, with $e_k \in H$ and $f_k \in V$ for all k. We write $B_H = (e_1, \ldots, e_\ell)$ and $B_V = (f_1, \ldots, f_\ell)$. We consider a symplectic isomorphism ψ with matrix $\mathbf{P}$ relative to B:

$$\mathbf{P} = \begin{bmatrix} \mathbf{A} & \mathbf{C} \\ \mathbf{B} & \mathbf{D} \end{bmatrix}$$

the symplectic character of ψ being equivalent to the identities

$${}^t\mathbf{AB} = {}^t\mathbf{BA}, \quad {}^t\mathbf{DC} = {}^t\mathbf{CD}, \quad {}^t\mathbf{DA} - {}^t\mathbf{CB} = \mathbf{I}.$$

Given a Lagrangian plane L with coordinate γ, and the matrix $\mathbf{\Gamma}$ of γ relative to B_H, we look for the coordinate γ' of the image $L' = \psi(L)$, which here we assume to be transverse to V. If l is the natural embedding of L, the matrix relative to B_H and B of the composition $\psi \circ l$, which maps H onto L', reads:

$$\begin{bmatrix} \mathbf{A} + \mathbf{C}\mathbf{\Gamma} \\ \mathbf{B} + \mathbf{D}\mathbf{\Gamma} \end{bmatrix},$$

and one gets easily the desired formula for the matrix $\mathbf{\Gamma}'$ of γ' in B_H, namely:

PROPOSITION. *With the notation as above, the variance of the coordinate is given by*

$$(V\Gamma) \qquad \mathbf{\Gamma}' = (\mathbf{B} + \mathbf{D}\mathbf{\Gamma})(\mathbf{A} + \mathbf{C}\mathbf{\Gamma})^{-1}$$

where $\mathbf{\Gamma}$ is the coordinate of L and $\mathbf{\Gamma}'$ the coordinate of the transformed space $\psi(L)$.

Note that the invertibility of $\mathbf{A} + \mathbf{C}\mathbf{\Gamma}$ is equivalent to the transversality of $\psi(L)$ to V and that $\mathbf{\Gamma}'$ is *symmetric* due to the symplectic character of $\mathbf{P}$. Of course $\mathbf{\Gamma}$ and $\mathbf{\Gamma}'$ are also twice the matrices associated with the generating functions of L and L'.

1.5.2.
We are now in a convenient position to study the effect of symplectic transformations on the notions introduced above. We start with symplectic angular forms. So let again ψ be a symplectic isomorphism of $\mathbb{R}^{2\ell}$, and let l_1, l_2 be two linear Lagrangian embeddings. The following is obvious:

PROPOSITION. *The maps $l'_i = \psi \circ l_i$, $i = 1, 2$, are Lagrangian embeddings, and the symplectic angular form of the pair (l'_1, l'_2) coincides with that of (l_1, l_2), i.e. the symplectic angular form is a symplectic invariant.*

1.5.3. Let us now examine the more interesting case of splitting forms. So let H and V be transverse Lagrangian spaces and let $L_1, L_2 \in \Lambda_V$. As one would expect the splitting form relative to H and V transforms quite simply under the symplectic isomorphism ψ.

PROPOSITION. *Let a and b be in* H, *and denote the images under ψ with primed letters. Then:*
$$\big(\sigma_{H'V'}(L'_1, L'_2)\big)(a', b') = \big(\sigma_{HV}(L_1, L_2)\big)(a, b).$$
The splitting form is thus equivariant *under the action of the symplectic group.*

PROOF. One checks that for $k = 1, 2$ the natural embeddings $l'_{L'_k}$ and l_{L_k} associated with L'_k and L_k relative to the pairs (H', V') and (H, V) respectively are conjugate:
$$l_{L'_k} = \psi \circ l_{L_k} \circ (\psi^{-1})_{|H'}.$$
Thus, writing σ and σ' for $\sigma_{HV}(L_1, L_2)$ and $\sigma_{H'V'}(L'_1, L'_2)$ one gets
$$\sigma'(a', b') = \Omega\Big(\psi \circ l_{L_1} \circ (\psi^{-1})(a'), \psi \circ l_{L_2} \circ (\psi^{-1})(b')\Big) = \Omega\big(l_{L_1}(a), l_{L_2}(b)\big) = \sigma(a, b).$$
$\square$

1.5.4. The variance formula for the splitting matrix. Note that in the preceding formula the image of the splitting form is evaluated relatively to the *images* H' and V' under ψ of the initial reference subspaces. The next step is to describe the variance of the splitting form of two Lagrangian spaces, relative to a *fixed* pair (H, V), which we do explicitly, using matrices. It is of course possible to compute the coordinates of L_1 and L_2 using the formulas in 1.5.1 and then to take their difference, but there is a shortcut. Writing $[\Omega]$ for the matrix of the symplectic form Ω relative to B, one can write:
$$\mathbf{\Gamma}_2 - \mathbf{\Gamma}_1 = {}^t\mathbf{\Gamma}_2 - \mathbf{\Gamma}_1 = [\mathbf{I} \ {}^t\mathbf{\Gamma}_2] \, [\Omega] \begin{bmatrix} \mathbf{I} \\ \mathbf{\Gamma}_1 \end{bmatrix} = [\mathbf{I} \ {}^t\mathbf{\Gamma}_2] \, {}^t\mathbf{P} \, [\Omega] \, \mathbf{P} \begin{bmatrix} \mathbf{I} \\ \mathbf{\Gamma}_1 \end{bmatrix}$$
and the last term of this equality reads
$$[{}^t(\mathbf{A} + \mathbf{C}\mathbf{\Gamma}_2) \ {}^t(\mathbf{B} + \mathbf{D}\mathbf{\Gamma}_2)] \, [\Omega] \begin{bmatrix} \mathbf{A} + \mathbf{C}\mathbf{\Gamma}_1 \\ \mathbf{B} + \mathbf{D}\mathbf{\Gamma}_1 \end{bmatrix} = {}^t(\mathbf{A} + \mathbf{C}\mathbf{\Gamma}_2)({}^t\mathbf{\Gamma}'_2 - \mathbf{\Gamma}'_1)(\mathbf{A} + \mathbf{C}\mathbf{\Gamma}_1)$$
where $\mathbf{\Gamma}'_k = (\mathbf{B} + \mathbf{D}\mathbf{\Gamma}_k)(\mathbf{A} + \mathbf{C}\mathbf{\Gamma}_k)^{-1}$, $k = 1, 2$. Inverting this last equation, one gets the

PROPOSITION. *With the notations as above, the variance of the splitting matrix is given by:*

(VS) $$\mathbf{S}' = {}^t(\mathbf{A} + \mathbf{C}\mathbf{\Gamma}_2)^{-1} \, \mathbf{S} \, (\mathbf{A} + \mathbf{C}\mathbf{\Gamma}_1)^{-1}$$

where $\mathbf{S} = \mathbf{\Gamma}_2 - \mathbf{\Gamma}_1$ *and* $\mathbf{S}'$ *are the splitting matrices of the pairs* (L_1, L_2) *and* (L'_1, L'_2) *respectively.*

This variance equality (VS) which is straightforward from the definition of the splitting matrix will play a crucial role in Chapter 2, when we draw the consequences in a perturbative setting. Let us stress again that this formula essentially enables us to keep track of the splitting matrix and other related objects after performing (normalizing) canonical transformations. Note that due to the symplectic character of the matrix **P** one needs to know only "half" of it in order to compute its effect on the splitting form. But it is also important to remark that (VS) is *not* a closed formula for the transformation of the splitting matrix, since it contains both matrices $\mathbf{\Gamma}_1$ and $\mathbf{\Gamma}_2$. This is in line with previous remarks and indeed there cannot exist a closed formula for the transformation of the splitting matrix.

1.5.5. Sections and coordinates. In Chapters 2 and 3 we will be especially concerned with the problem of evaluating splitting forms attached to pairs of Lagrangian submanifolds in the same energy level of a Hamiltonian system on $T^*\mathbb{T}^\ell$. Since these manifolds are invariant under the Hamiltonian flow, their intersection is also invariant and (in general) is the union of nontrivial orbits; the situation is thus necessarily degenerate. A natural method to bypass this problem is to introduce a section in phase space, transverse to the vector field, and then to consider the intersections of the submanifolds with that section. It is thus useful to understand the relation between the splitting form one gets for two different choices of sections.

This is of course a *linear* problem, and we therefore limit ourselves to the case of pairs of Lagrangian planes in $\mathbb{R}^{2\ell}$ (with the standard form Ω); we denote as usual by H_0 and V_0 the horizontal and vertical spaces in $\mathbb{R}^{2\ell}$. So we start with a given hyperplane $\mathbf{H}$ (the tangent space at some point of the energy level) and we consider the splitting forms of pairs of Lagrangian spaces (the tangent spaces to the Lagrangian submanifolds) which are *contained* in $\mathbf{H}$. Note first that $\mathbf{H}$ is a coisotropic subspace of $\mathbb{R}^{2\ell}$, meaning that the (one-dimensional) subspace $X = \mathbf{H}^\perp$ is contained in $\mathbf{H}$; of course if L is a Lagrangian space contained in $\mathbf{H}$, X is contained in L, since $X = \mathbf{H}^\perp \subset \mathrm{L}^\perp = \mathrm{L}$, and this is why we want to make use of sections in this problem. We may assume that X is not contained in V_0 (this condition will always be met in our setting), or equivalently that $\mathbf{H}$ does not contain the Lagrangian space V_0; the intersection $\mathbf{V} = \mathrm{V}_0 \cap \mathbf{H}$ is thus $(\ell - 1)$-dimensional, and X has a one-dimensional projection on the horizontal space H_0, which we denote by ξ.

Let us now introduce the sections. We will consider only those which are obtained as the intersection of $\mathbf{H}$ with hyperplanes of the form $\widehat{C} = C \oplus \mathrm{V}_0$, where C is a given hyperplane of H_0 *transverse to* ξ, and we will write $\Sigma_C = \mathbf{H} \cap \widehat{C}$. The transversality condition immediately implies that $\widehat{C}$ is transverse to $\mathbf{H}$, and so the intersection Σ_C is $2(\ell-1)$-dimensional. One sees easily that Σ_C is indeed *symplectic* since it is transverse to X and contained in $\mathbf{H}$. One remarks furthermore that if $(e_1, \ldots, e_\ell)$ is some basis of H_0, with $(e_1, \ldots, e_{\ell-1})$ contained in C and $e_\ell \in \xi$, then the unique basis $(f_1, \ldots, f_\ell)$ of V_0 such that (e_i, f_j) is symplectic satisfies $f_i \in \mathbf{V}$, $1 \leq i \leq \ell - 1$.

The main point that we wish to emphasize is that in general the horizontal space H_0 is *not* contained in $\mathbf{H}$, and thus in order to get a horizontal space inside $\mathbf{H}$ one has to project the set C on $\mathbf{H}$, *e.g.* along f_l. This construction is not canonical and the result depends in all cases on the choice of C: there is little to add in the general case; one simply has to introduce suitable bases and use the transformation formulas given in the preceding paragraphs.

Recall now that we are only concerned with *splitting forms*, and not with *coordinates* of Lagrangian spaces in $\mathbf{H}$, and that these forms are the *differences* of the coordinates attached to pairs of Lagrangian spaces contained in $\mathbf{H}$. This amounts in fact to considering that the kernel X is contained in the horizontal space, or equivalently that $\mathbf{H}$ contains H_0, the problem being now to compare the coordinate of some space $\mathrm{L} \in \mathbf{H}$ relative to H_0 and V_0 with the coordinate relative to C and $\mathbf{V}$

of the intersection $L_C = L \cap \Sigma_C$. So let us denote by j and γ the equation and coordinate of L relative to H_0, V_0, viz. $L = \{x + j(x), \ x \in H_0\}$. One sees immediately that:
$$L_C = \{x + j(x), \ x \in C\},$$
and so the equation of L_C relative to C and $\mathbf{V}$ is the restriction $j_C = j_{|C}$; note that j takes its values in $\mathbf{V}$ since $j(x) = (x + j(x)) - x \in \mathbf{H}$ if $x \in H_0$. We denote by γ_C the restriction of γ to C, and see in the same way that γ_C is the coordinate of L_C relative to C and $\mathbf{V}$.

We now compare the spectral elements of γ and its restriction γ_C, and we assume now that the horizontal space H_0 is endowed with the usual Euclidean metric. Note that ξ is contained in the kernel of γ, and hence is an orthogonal direct factor for γ. A particularly simple case arises when $C = \xi^\perp$ (Euclidean orthogonal): the eigenspaces of the restriction γ_C are eigenspaces for γ, and the associated eigenvalues coincide.

In general the dynamical features of our problem will not allow us to choose such simple sections. For a general hyperplane C transverse to ξ, we introduce the projection π from $\xi^\perp$ onto C, parallel to ξ. Since ξ is contained in the kernel of γ, one sees that
$$\gamma(x, y) = \gamma(\pi(x), \pi(y))$$
for all $x, y \in C$. Denoting by B_C and $B_\perp$ two orthogonal bases in C and $\xi^\perp$, the matrices $\mathbf{\Gamma}_C$ and $\mathbf{\Gamma}_\perp$ of γ_C and of the restriction $\gamma_\perp$ to $\xi^\perp$ in B_C and $B_\perp$ are thus related by an equation of the form

(VΣ) $$\mathbf{\Gamma}_C = {}^t\mathbf{P}\,\mathbf{\Gamma}_\perp\,\mathbf{P},$$

where $\mathbf{P}$ is the matrix of π in B_C and $B_\perp$. We will see in §2.3.2 that such a relation fits our needs, basically because we are dealing with *symmetric* matrices, so that one can satisfactorily compare the spectral elements of $\mathbf{\Gamma}_C$ and $\mathbf{\Gamma}_\perp$, at least in the perturbative setting and as far as asymptotic properties with respect to some perturbation parameter are concerned.

1.5.6. From coordinates to Euclidean angles. We now consider two Lagrangian spaces L_1, L_2 transverse to V_0, together with their coordinates γ_1, γ_2 relative to H_0 and V_0, and ask for their Euclidean angle. The question was already answered in Proposition 1.4.1 in the case when $L_1 = H_0$ and $L_2 = L \in \Lambda_{V_0}$: then the eigenspaces of the representative $\varphi_{H_0,L}$ are the same as those of γ, and if α is an eigenvalue of γ, the corresponding eigenvalue for $\varphi_{H_0,L}$ is $\rho = (1 + i\alpha)/|1 + i\alpha|$.

The general case is reduced to the preceding situation by introducing the unitary morphism $\varphi_{H_0 L_1}$ which maps H_0 onto L_1 and considering the subspace $L = \varphi_{H_0 L_1}^{-1}(L_2)$. Since $\varphi_{H_0 L_1}^{-1}$ is unitary, the pairs (H_0, L) and (L_1, L_2) define the same Euclidean angle, and since L is transverse to V_0 one can apply Proposition 1.4.1 to get the determination of the angle (H_0, L). Writing as in that proposition $\varphi_{H_0 L_1} = (I + i J_1) \circ R_1^{-1}$ and using bold letters for matrices in a *unitary* basis of H_0, we find the following expression for the matrix of the equation of L:

$$\mathbf{\Gamma}' = {}^t(\mathbf{I} + \mathbf{\Gamma}_1 \mathbf{\Gamma}_2)^{-1} \mathbf{R}_1 \,(\mathbf{\Gamma}_2 - \mathbf{\Gamma}_1)\, \mathbf{R}_1 \,(\mathbf{I} + \mathbf{\Gamma}_1^2)^{-1}$$

whose spectral elements give the measure (and a determination) of the Euclidean angle. Note that this formula is nothing else but the variance equality for the splitting matrix (see Proposition 1.5.4) supplemented with the additional information that the morphism is unitary.

The variance problem for the Euclidean angles is more complicated. Let us first make a general remark about the polar decomposition of symplectic isomorphisms. It is well-known that a symplectic isomorphism ψ (or simply a real isomorphism of $\mathbb{R}^{2\ell}$) can be uniquely written in the form $\psi = s \circ \varphi$, where $\varphi \in O(2\ell, \mathbb{R})$ and s is a symmetric positive definite linear map. Obviously $\psi \circ {}^t\psi = (s \circ \varphi) \circ {}^t(s \circ \varphi) = s^2$, and thus s is the (unique) positive square root of $\psi \circ {}^t\psi$. It is easy to verify that s is *symplectic*, and thus that $\varphi \in Sp(2\ell, \mathbb{R}) \cap O(2\ell, \mathbb{R}) = U(\ell, \mathbb{C})$. The classical polar decomposition thus provides a homeomorphism (not a homomorphism of course) between $Sp(2\ell, \mathbb{R})$ and $U(\ell) \times \mathcal{S}(2\ell)$, where $\mathcal{S}(2\ell)$ is the set of symplectic symmetric positive definite linear isomorphisms of $\mathbb{R}^{2\ell}$. Since the isomorphism φ is orthogonal, the effect of a symplectic isomorphism ψ on the Euclidean angle defined by a pair (L_1, L_2) reduces to that of an element $s \in \mathcal{S}(2\ell)$. Now, for an *explicit* determination of the variance, we will always use coordinates as an intermediate step, as they are easier to evaluate: we first compute the coordinates relative to a suitable decomposition, then use the preceding transformation formula to get the new coordinates and finally deduce the new Euclidean angles.

We conclude this paragraph with a general remark related to the above transformation formulas: in spite of its dependence on the choice of a horizontal/vertical decomposition of the ambient space, the splitting form appears as the most natural object to handle from the computational viewpoint, since its evaluation and transformation rule are quite straighforward. By contrast the Euclidean angle encodes the geometric information more directly and is thus better adapted to formal descriptions, such as the persistence of hyperbolicity under perturbation. We recall that in Chapters 2 and 3 we will be mainly concerned with the evaluation of the splitting form in a perturbative setting.

1.6. The splitting of Lagrangian submanifolds

In this paragraph we give the global form of the preceding linear notions. Athough this should definitely be written up for completeness, much of it is actually routine and does entail rather heavy notation. Moreover it is too general for the perturbative setting which will essentially concern us in further chapters so that skipping this paragraph at first reading is quite possible and perhaps advisable. But it does give a better geometric understanding of the situation and should certainly be kept in store in view of more global and geometric investigations.

To start with we consider pairs of Lagrangian submanifolds of a given symplectic manifold (M, Ω) and globalize the various definitions of "splitting". We need first to assume some intersection property and some regularity for the intersection: if $\mathcal{L}_i : \mathbf{L}_i \to M$, $i = 1, 2$, are two injective Lagrangian immersions, we say that the pair $(\mathcal{L}_1, \mathcal{L}_2)$ is *regular* when the intersection $I = \mathcal{L}_1(\mathbf{L}_1) \cap \mathcal{L}_2(\mathbf{L}_2)$ is nonempty, the sets $\mathbf{I}_i = \mathcal{L}_i^{-1}(I)$ are submanifolds of $\mathbf{L}_i$, and the restrictions of $\mathcal{L}_i$ to $\mathbf{I}_i$ are immersions. In that case, we consider the set $\mathbf{I}$ obtained from the disjoint union $\mathbf{I}_1 \coprod \mathbf{I}_2$ by identifying the pairs of points having the same images under $\mathcal{L}_1$ and $\mathcal{L}_2$, so one can consider $\mathbf{I}$ as a submanifold of both $\mathbf{L}_1$ and $\mathbf{L}_2$, and we define naturally the map $\mathcal{I} : \mathbf{I} \to M$ as the restriction of $\mathcal{L}_i$ to $\mathbf{I}_i$; we say (somewhat improperly) that $\mathcal{I}$ is the *intersection* of $\mathcal{L}_1$ and $\mathcal{L}_2$.

Of course two Lagrangian immersions or submanifolds transverse in M and with nonempty intersection define a regular pair, but due to dimensions the intersection is then simply a union of isolated points. A more interesting case is the one of submanifolds of the same (regular) level $\mathcal{H}$ of a Hamiltonian function on M; the transversality relative to $\mathcal{H}$ implies that the intersections are one-dimensional, being thus the union of isolated orbits of the Hamiltonian vector field (recall that the Lagrangian submanifolds of $\mathcal{H}$ are automatically invariant under the Hamiltonian flow). But non-transverse cases will also occur naturally in the sequel, in connection with the invariant manifolds of invariant tori near multiple resonances.

1.6.1. Splitting forms and Euclidean angles of pairs of Lagrangian submanifolds. This paragraph is devoted to straightforward but necessary translations from the linear case to a coordinate free setting. In §1.7 below we specialize to the case when M is a cotangent bundle, which is the most important example and makes it possible to write down—some—formulas.

So for the time being we begin quite generally with Lagrangian immersions: let $(\mathcal{L}_1, \mathcal{L}_2)$ be a regular pair of injective Lagrangian immersions $\mathbf{L}_i \to M$, and denote by $\mathcal{I}$ their intersection. Since $\mathbf{I} \subset \mathbf{L}_i$ for $i = 1, 2$, we can define the two vector bundles $T_{\mathbf{I}}\mathbf{L}_i$, and consider their product $\mathbf{X}$ over $\mathbf{I}$, whose fiber over $x = (x_1, x_2)$ is the product $T_{x_1}\mathbf{L}_1 \times T_{x_2}\mathbf{L}_2$. We consider finally the bundle $\mathcal{B}(\mathbf{X})$ of bilinear forms of $\mathbf{X}$, whose fiber over x is $\mathrm{Bil}(T_{x_1}\mathbf{L}_1, T_{x_2}\mathbf{L}_2)$. Note that given $x = (x_1, x_2) \in \mathbf{I}$, the maps $T_{x_i}\mathcal{L}_i$ are Lagrangian linear embeddings of the spaces $T_{x_i}\mathbf{L}_i$ into $T_{\mathcal{I}(x)}M$.

DEFINITION. The symplectic angular form of the pair $(\mathcal{L}_1, \mathcal{L}_2)$ is the section $\mathcal{A}$ of the bundle $\mathcal{B}(\mathbf{X})$ whose value over $x = (x_1, x_2) \in \mathbf{I}$ is the symplectic angular form $\mathcal{A}$ associated with the pair $(T_{x_1}\mathcal{L}_1, T_{x_2}\mathcal{L}_2)$, i.e. for $a \in T_{x_1}\mathbf{L}_1$ and $b \in T_{x_2}\mathbf{L}_2$:

$$\mathcal{A}_x(a, b) = \Omega_{\mathcal{I}(x)}\Big(T_{x_1}\mathcal{L}_1(a), T_{x_2}\mathcal{L}_2(b)\Big).$$

One has the same invariance property as in the linear case, to wit:

PROPOSITION. *If Φ is a symplectic diffeomorphism of M, the maps $\mathcal{L}'_i = \Phi \circ \mathcal{L}_i$ define a regular pair of Lagrangian immersions with the same intersection set $\mathbf{I}$, and the symplectic angular form of the pair $(\mathcal{L}'_1, \mathcal{L}'_2)$ is equal to the one of $(\mathcal{L}_1, \mathcal{L}_2)$, i.e. the symplectic angular form is invariant under symplectic diffeomorphisms.*

1.6.2. Let us move on to global splitting forms, assuming that we are given in the tangent bundle TM two supplementary Lagrangian subbundles $\mathcal{H}$ and $\mathcal{V}$ (i.e. such that the Whitney sum $\mathcal{H} \oplus \mathcal{V}$ is TM); denote by $p_{\mathcal{H}}$, $p_{\mathcal{V}}$ the associated projections from TM onto $\mathcal{H}$ and $\mathcal{V}$. Let now Λ be the set of all Lagrangian submanifolds of M and $\Lambda_{\mathcal{V}}$ the set of Lagrangian submanifolds everywhere transverse to $\mathcal{V}$. Proceeding as in the linear case again, we denote by $\mathcal{S}(\mathcal{H})$ the bundle of symmetric bilinear forms of the subbundle $\mathcal{H}$, and for every submanifold $\mathcal{I}$ of M, $\mathcal{S}_{\mathcal{I}}(\mathcal{H})$ the restriction to $\mathcal{I}$ of $\mathcal{S}(\mathcal{H})$. Remark that if $\mathcal{L}$ is in $\Lambda_{\mathcal{V}}$ and $x \in \mathcal{L}$, the tangent space $T_x\mathcal{L}$ is a Lagrangian subspace of the symplectic vector space (T_xM, Ω_x), transverse to $\mathcal{V}_x$ and one can define as in the linear case the natural linear embedding $l_x^{\mathcal{L}}$ of $T_x\mathcal{L}$ associated with the decomposition $(\mathcal{H}_x, \mathcal{V}_x)$ of T_xM. We are thus led to the

DEFINITION. Let $(\mathcal{L}_1, \mathcal{L}_2)$ be a regular pair of Lagrangian submanifolds of M, with $\mathcal{L}_i \in \Lambda_{\mathcal{V}}$ for $i = 1, 2$, and $\mathcal{I} = \mathcal{L}_1 \cap \mathcal{L}_2$. The splitting form of the pair $(\mathcal{L}_1, \mathcal{L}_2)$, relative to $\mathcal{H}, \mathcal{V}$, is the section Σ of $\mathcal{S}_{\mathcal{I}}(\mathcal{H})$ whose value over $x \in \mathcal{I}$ is the splitting form $\sigma_{\mathcal{H}_x, \mathcal{V}_x}$ of the pair of linear Lagrangian embeddings $(l_x^{\mathcal{L}_1}, l_x^{\mathcal{L}_2})$.

An analogous definition can be formulated for pairs of Lagrangian immersions. Clearly the same equivariance property as in the linear case holds:

PROPOSITION. *Let Φ be a symplectic diffeomorphism of M, and $(\mathcal{L}_1, \mathcal{L}_2)$ a regular pair of Lagrangian submanifolds of M, with $\mathcal{L}_i \in \Lambda_\mathcal{V}$. Let us denote by primed letters the images by Φ. Then for $x \in \mathcal{I}$ and $x' = \Phi(x) \in \mathcal{I}'$, the splitting form $\Sigma'_{x'}$ of the images $(\mathcal{L}'_1, \mathcal{L}'_2)$, relative to the decomposition $(\mathcal{H}', \mathcal{V}')$, is the image under $T_x\Phi$ of the splitting form Σ_x relative to the decomposition $(\mathcal{H}, \mathcal{V})$:*

$$\Sigma'_{\Phi(x)}\Big(T_x\Phi(a), T_x\Phi(b)\Big) = \Sigma_x(a,b).$$

The symplectic angular form is equivariant under symplectic diffeomorphisms.

One then easily translates the linear transformation formulas for the splitting form under symplectomorphisms (see §1.5).

1.6.3. Let us finally state the definitions for Euclidean angles. We assume that the symplectic manifold (M, Ω) is endowed with a compatible almost-complex structure, *i.e.* an operator J of TM such that $J^2 = -\operatorname{Id}$ and such that the bilinear map $(a, b) \mapsto \Pi(a, b) = \Omega(J(a), b)$ is positive definite in each fiber of TM, and defines thus a Riemannian metric on M. In that case TM has a Hermitian structure given by $\Xi(a, b) = \Pi(a, b) + i\Omega(a, b)$, and we can speak of the orthogonal group bundles $\mathcal{U}$ and $\mathcal{O}$ whose fibers over $x \in M$ are $U(T_xM, \Xi)$ and $O(T_xM, \Pi)$. For $x \in M$, it is thus possible to define the space $\mathcal{E}_x$ of Euclidean angles of Lagrangian subspaces of T_xM, we write $\mathcal{E}$ for the bundle (over M) of Euclidean angles and $\mathcal{E}_\mathcal{I}$ for its restriction to a submanifold $\mathcal{I}$. This being said we can phrase the

DEFINITION. *Let $(\mathcal{L}_1, \mathcal{L}_2)$ be a regular pair of Lagrangian submanifolds of M, with $\mathcal{I} = \mathcal{L}_1 \cap \mathcal{L}_2$. The Euclidean angle of $(\mathcal{L}_1, \mathcal{L}_2)$ is the section of $\mathcal{E}_\mathcal{I}$ whose value at $x \in \mathcal{I}$ is the Euclidean angle of $(T_x\mathcal{L}_1, T_x\mathcal{L}_2)$. One defines in a natural way the representative of the angle, as the section of $\mathcal{U}_\mathcal{I}$ whose value over $x \in \mathcal{I}$ is the representative of $(T_x\mathcal{L}_1, T_x\mathcal{L}_2)$, and the associated sections given by the measures and the determinations.*

Note finally that one can pass from coordinates to Euclidean angles in the same way as in the linear case.

1.7. Lagrangian submanifolds in a cotangent bundle

We now study more concretely the case of a cotangent bundle $(M = T^*V, \pi, V)$, endowed with its canonical Liouville exact symplectic structure. One can keep in mind the primary example of the ℓ-dimensional torus $V = \mathbb{T}^\ell = \mathbb{R}^\ell/\mathbb{Z}^\ell$ (or an open subset of it). This example is however not quite generic because then the bundle T^*V (as well as TV) is *canonically* identified with $V \times \mathbb{R}^\ell$. As a consequence $T(T^*V)$ is just the product $(V \times \mathbb{R}^\ell) \times (\mathbb{R}^\ell \times \mathbb{R}^\ell)$, and we have a *canonical* horizontal/vertical decomposition $T(T^*V) = \mathcal{H}_0 \oplus \mathcal{V}_0$. For general V a *vertical* subbundle can always be defined intrinsically as the subspace of all vectors tangent to the fibers T_x^*V, but the horizontal one cannot and depends on (in fact is equivalent to) the choice of a connection on V.

1.7.1. Cohomology and defect of exactness. We restrict ourselves to submanifolds defined as images of one-forms on V. In the case of a trivial bundle, these are graphs of maps from V to $\mathbb{R}^\ell$, so that in all cases we say that such submanifolds have the *graph property*. If $\mathcal{L} = \operatorname{im} \beta$, the form β defines an embedding of V into T^*V with image $\mathcal{L}$, hence an identification between V and $\mathcal{L}$, and it makes sense to speak of the pull-back $\beta^*(\lambda)$ which is a one-form on V and in fact is just equal to β by the characterization of the Liouville form λ (*cf.* §1.1). Thus $\mathcal{L}$ is Lagrangian if and only if β is closed, and exact if and only if β is. In this last case, the manifold is the graph of the derivative of a function $S : V \to \mathbb{R}$ which is called a *generating function* for $\mathcal{L}$.

More generally, when $\mathcal{L}$ is Lagrangian, we say that the cohomology class $\delta \in H^1(V, \mathbb{R})$ of β is the *defect of exactness* of $\mathcal{L}$. In the case $V = \mathbb{T}^\ell$, the vector space $H^1(\mathbb{T}^\ell, \mathbb{R})$ is ℓ-dimensional. In fact using the canonical coordinate system $(\theta_1, \ldots, \theta_\ell)$ on $\mathbb{R}^\ell$ (the universal cover of $\mathbb{T}^\ell$), the forms $d\theta_1, \ldots, d\theta_\ell$ (or rather their cohomology classes) form a natural basis for $H^1(\mathbb{T}^\ell, \mathbb{R})$, so that $\delta \in \mathbb{R}^\ell$ is an ℓ-vector. If we lift (canonically) the coordinate system (θ_i) to a coordinate system (θ_i, z_i) of $T^*\mathbb{T}^\ell = \mathbb{T}^\ell \times \mathbb{R}^\ell$ (of the universal cover rather), then z_i is just the component on $d\theta_i$, and the defect of exactness appears as (minus) the *translation* in the fibers one has to operate in order to pass from the initial Lagrangian manifold to an exact one.

1.7.2. Expressions in local coordinates. Before proceeding, we insert a word of caution about notation: the letter S is used above to comply with a long tradition (dating back from the 18-th century) to denote the action integral. So S certainly does *not* stand for "splitting". Referring to §1.3.1, in the linear case one has $S = Q_{\mathrm{L}}$ which is a quadratic form. So $S = Q_{\mathrm{L}}$ computes the coordinate γ_{L} as will also be the case below in the global situation. It is of course the same S which occurs in Chapters 2 and 3 in a perturbative setting, where the splitting is determined by a difference $S^+ - S^-$. More precisely, at an intersection point the splitting matrix is given by the difference of the Hessian matrices (or the Hessian matrix of the difference) of two generating functions (S^+ and S^- in Chapters 2 and 3) as is also apparent below. As a rule, we use boldscript to denote splitting matrices and plainscript for the generating functions.

With this in mind we return to the effective "computation" of the various objects. Let (x_i) be a coordinate system over an open subset $U \subset V$ and let (x_i, y_i) be the associated coordinate system on $U^* = \pi^{-1}(U) \subset T^*V$, the coordinates (y_i) being thus the components on the basis of one-forms (dx_i). Let ϖ denote the natural projection $T(T^*V) \to T^*V$. We can lift again the preceding coordinates and obtain a coordinate system $(x_i, y_i, \xi_i, \eta_i)$ on $\overline{U} = \varpi^{-1}(U^*)$, the coordinate ξ_i (resp. η_i) being the component on the vector field $\partial/\partial x_i$ (resp. $\partial/\partial y_i$). Note that the basis $(\partial/\partial x_i, \partial/\partial y_i)$ of $T_{(x,y)}T^*V$ is symplectic for the canonical symplectic form $\Omega_{(x,y)}$. Note furthermore that over the open set $\overline{U}$ the vector fields $(\partial/\partial x_i)$ and $(\partial/\partial y_i)$ generate two supplementary horizontal and vertical subbundles $\mathcal{H}$ and $\mathcal{V}$, so that one can speak of the coordinates and splitting forms relative to that decomposition. Finally $\mathcal{V}$ is always the restriction to U^* of the vertical subbundle of $T(T^*V)$, whereas $\mathcal{H}$ depends on the initial choice of local coordinates.

Given $\mathcal{L} = \mathrm{im}\,\beta \subset T^*V$ a Lagrangian submanifold and x a point of V, we first determine the tangent space $T_{\beta(x)}\mathcal{L}$. As a first step, since β is an embedding of V onto $\mathcal{L}$, one gets directly $T_{\beta(x)}\mathcal{L} = T_x\beta(T_xV)$. Consider now a local coordinate system (x_i) with lift $(x_i, y_i, \xi_i, \eta_i)$ to $T(T^*V)$ and write $\beta = (\beta_i)$ in these coordinates. The tangent map $T\beta : TV \to T(T^*V)$ reads:

$$(x_i, \xi_i) \mapsto \left(x_i,\; \beta_i(x),\; \xi_i,\; \sum_{j=1}^n \frac{\partial \beta_i}{\partial x_j}(x)\,\xi_j\right),$$

and the subspace $T_{\beta(x)}\mathcal{L}$ appears as the graph of the linear map

$$\xi \mapsto \sum_{j=1}^{n} \left(\partial \beta_i / \partial x_j(x) \right) \xi_j$$

from $T_x V$ to the vertical fiber of $T_{\beta(x)}(T^*V)$. Let $\mathcal{H}$ and $\mathcal{V}$ be as above; since the basis $(\partial/\partial x_i, \partial/\partial y_i)$ is symplectic, we see that in the basis $(\partial/\partial x_i)$, the matrix $\mathbf{\Gamma}_x$ of the coordinate γ_x of the Lagrangian subspace $T_{\beta(x)}\mathcal{L}$ relative to $\mathcal{H}$ and $\mathcal{V}$ is just:

$$\mathbf{\Gamma}_x = \left[\frac{\partial \beta_i}{\partial x_j}(x) \right].$$

Particularizing to the case of an exact submanifold $\mathcal{L}_S = \operatorname{im} dS$, we say (improperly) that the map $TdS : TV \to T(T^*V)$ is the *Hessian* of S (not to be confused with the second differential ddS, from TV to TTV). The tangent space $T_{dS(x)}\mathcal{L}_S$ is thus the image of the Hessian of S at x, and the matrix of its coordinate in the preceding basis is the usual Hessian matrix of S:

$$\mathbf{\Gamma}_x = \left[\frac{\partial^2 S}{\partial x_i \partial x_j}(x) \right].$$

Given now *two* Lagrangian submanifolds $\mathcal{L}_i = \operatorname{im} \beta_i$, $i = 1, 2$, an intersection point $\chi \in \mathcal{L}_1 \cap \mathcal{L}_2$, and a coordinate system around $x = \pi(\chi)$, we immediately get the splitting matrix of the pair of tangent spaces $(T_\chi \mathcal{L}_1, T_\chi \mathcal{L}_2)$ in the lifted coordinate system: just replace β by $\beta_2 - \beta_1$ in the formula above. In the case of exact submanifolds the same is true, replacing S by $S_2 - S_1$, which is the first step in the Hamilton-Jacobi method of Chapter 3.

1.7.3. The effect of exact symplectomorphisms.

We finally have to (briefly) consider the transformation of coordinates and splitting forms under symplectomorphisms. In the case of a general symplectomorphism of T^*V, there is nothing more to say, but we will examine more particularly the exact symplectomorphisms obtained as lifts of diffeomorphisms of the "configuration space" V. Their characterization among all the fibered diffeomorphisms of T^*V and their effect on exact manifolds are given in the following general proposition, whose proof is straightforward.

PROPOSITION. *Let $f : W \to V$ a diffeomorphism between two manifolds. Consider a map $\Phi : T^*W \to T^*V$ fibered over f, i.e. such that the following diagram commutes:*

$$\begin{array}{ccc} T^*W & \xrightarrow{\Phi} & T^*V \\ \pi_W \downarrow & & \downarrow \pi_V \\ W & \xrightarrow{f} & V \end{array}$$

Then one has the following properties:
— Φ is exact symplectic if and only if there exists a function $F : W \to \mathbb{R}$ such that

$$\Phi(\chi) = [{}^t(T_x f)]^{-1}(\chi + dF(x)),$$

in which case we write $\Phi = \Phi_f^F$. The inverse of Φ_f^F is $[\Phi_f^F]^{-1} = \Phi_{f^{-1}}^{-F \circ f}$.
— If $\mathcal{L}_S$ is exact with generating function S, the image $\Phi_f^F(\mathcal{L}_S)$ is exact with generating function $(S + F) \circ f$.

The role of the function F is simply to translate in the fibers by a covector $dF(x)$. It may be convenient to set $F = 0$, but other choices are useful. In particular an exact submanifold $\operatorname{im} dS$ can always be mapped to the zero section by means of an exact symplectomorphism, by choosing $f = id$ and $F = -S$. Here, since we are interested in the transformation formula for exact submanifolds, we can simply forget about F and confine ourselves to the class of cotangent lifts of diffeomorphisms which are linear on the fibers. So we consider an exact manifold $\widehat{\mathcal{L}} = \operatorname{im}(d\widehat{S})$ of W, and let $\Phi = \Phi_f^0$. The image $\Phi(\widehat{\mathcal{L}})$ has $S = \widehat{S} \circ f^{-1}$ as generating function, so that writing $g = f^{-1}$ for clarity one derives the expression (where $(\widehat{x}_i)$ and (x_i) are the coordinates systems in W and V respectively):

$$\frac{\partial^2 S}{\partial x_i \partial x_j}(x) = \sum_{k=1}^{\ell} \frac{\partial^2 g_k}{\partial x_i \partial x_j}(x) \frac{\partial \widehat{S}}{\partial \widehat{x}_k}(\widehat{x}) + \sum_{k,k'=1}^{\ell} \frac{\partial g_k}{\partial x_i}(x) \frac{\partial g_{k'}}{\partial x_j}(x) \frac{\partial^2 \widehat{S}}{\partial \widehat{x}_k \partial \widehat{x}_{k'}}(\widehat{x}),$$

from which one deduces the matrix of the coordinate γ of $T_\chi \mathcal{L}_S$, relative to the decomposition $(\mathcal{H}_\chi, \mathcal{V}_\chi)$ associated with the coordinates (x_i). Note that the first term involves the first derivatives $\frac{\partial \widehat{S}}{\partial \widehat{x}_k}(\widehat{x})$, which are nothing else but the coordinates of the point $\widehat{\chi}$, whereas the second term is the matrix expression of the direct image of the coordinate under the tangent map Tg. The first term is *not* tensorial in nature, whereas the second one is; this simply reflects the well-known non-tensorial character of the Hessian matrix at regular points.

As an application, we easily get the transformation formula for the splitting matrix at an intersection point of two exact Lagrangian submanifolds $\mathcal{L}_i = \operatorname{im}(dS_i)$, $i = 1, 2$. The matrix is given by:

$$\mathbf{S}(x) = \mathbf{\Gamma}_{2x} - \mathbf{\Gamma}_{1x} = \left[\frac{\partial^2(S_2 - S_1)}{\partial x_i \partial x_j}(x)\right],$$

which immediately yields the transformation rule:

$$\mathbf{S}_{ij}(x) = \sum_{k,k'=1}^{\ell} \frac{\partial g_k}{\partial x_i}(x) \frac{\partial g_{k'}}{\partial x_j}(x) \, \widehat{\mathbf{S}}_{kk'}(\widehat{x}),$$

which is of course again a tensorial formula. The above constitutes a computational proof of the

PROPOSITION. *With the assumptions and notation as above, the splitting form at the point χ, relative to $\mathcal{H}$ and $\mathcal{V}$, is the image under Φ of the splitting form at $\widehat{\chi}$, relative to $\widehat{\mathcal{H}}$ and $\widehat{\mathcal{V}}$.*

A direct proof is also immediate, since the tangent spaces to $\mathcal{L}$ at the point $\widehat{\chi}$ and to the image $\Phi(\mathcal{L})$ at the point $\Phi(\widehat{\chi})$ are obviously related via $T_{\widehat{\chi}}\Phi(T_{\widehat{\chi}}\mathcal{L}) = T_{\Phi(\widehat{\chi})}(\Phi(\mathcal{L}))$, whereas the horizontal and vertical subbundles correspond under the transformation Φ.

1.8. Hyperbolic tori and normally hyperbolic invariant manifolds

The model Hamiltonian $(*)$ of the introduction and more generally the normal forms in the neighborhood of resonances in the perturbative setting (to be studied in Chapter 2) have several very specific features, the main one for our concerns being the existence of *normally hyperbolic manifolds* containing the (partially hyperbolic) invariant tori. One knows also that there exist in general *non-KAM* invariant tori (*i.e.* which *cannot* be obtained by the Graff-Treschev method; see [H1], [Y1]) for which one cannot make use of the classical normal forms. In the present paragraph we first explore the symplectic geometry of the various hyperbolic invariant manifolds introduced in §1.2, showing that the dynamical hypotheses have indeed immediate and natural geometric consequences. We then introduce a new definition for hyperbolic tori in a purely dynamical setting, which is general enough to cover the case of non-KAM tori (for which one cannot assume the conjugacy of the flow on the torus to a linear rotation) and leads to the same conclusions regarding their symplectic geometry. In the next paragraph (§1.9) we take advantage of the geometric features of the invariant manifolds and establish general *straightening theorems*, which in turn make it possible to derive *geometric normal forms* for the invariant tori, without unnecessary arithmetical assumptions on their rotation vectors. Further extensions are possible and could be useful in order to take into account the existence of C^0 invariant sets, but such a generality will not be necessary here.

Let again (M, Ω) be a 2ℓ-dimensional smooth symplectic manifold, let $H \in C^r(M, \mathbb{R})$ $(2 \leq r \leq \omega)$ be a Hamiltonian on M, X_H the corresponding vector field, Φ its flow and ϕ the time-one map (note that we do not assume M exact). Here and below, "invariant" means "invariant under the flow Φ". In the sequel we make use of the definitions and notation introduced in §1.2. We will give only the ideas of the proofs, and refer to forthcoming work for full details.

1.8.1. Stable leaves of a pseudo-hyperbolic invariant manifold. We begin with the case of ρ-pseudo-hyperbolic manifolds, for which one can only prove (when $\rho < 1$) the existence of stable leaves attached to every point. Given two vectors tangent to a certain leaf at the same point, the symplectic character of Φ implies that their symplectic product remains constant under iteration, whereas their lengths tend to zero: one can thus expect that their symplectic product vanishes. This is indeed what our first statement asserts:

PROPOSITION. *Let V be a compact C^r ρ-pseudo-hyperbolic invariant submanifold of M, with $\rho \in\,]0,1[$, and assume that the stable bundle E^+ is C^1. Then for each $a \in V$, the ρ-stable leaf Δ_a^+ is isotropic.*

PROOF. Consider first the stable bundle E^+ and two vectors u, v in the same fiber E_p^+ $(p \in V)$. The symplectic character of ϕ leads to the equality $\Omega(u,v) = \Omega\big((T\phi)^n(u), (T\phi)^n(v)\big)$ for all $n \in \mathbb{Z}$, so that

$$\Omega(u,v) = \lim_{n \to +\infty} \Omega\big((T\phi)^n(u), (T\phi)^n(v)\big) = 0$$

since $\rho < 1$; this shows that E^+ is an isotropic *vector subbundle* of the tangent bundle of V.

The problem is now to pass from the linear objects to the nonlinear ones. We note first that by the *usual* tubular neighborhood theorem for compact submanifolds, one can find a C^1 diffeomorphism ν from a neighborhood U of V in M onto a neighborhood O of the zero section V_0 of a vector bundle (F, π, T) over V, such that $\nu(V) = V_0$. Identifying the objects in U and their images in O, we may assume furthermore that E^+ is a *vector* subbundle of F (see [Li-Ma]).

Fix on M a ρ-adapted Riemannian metric, and denote by $\| \, \|$ its image by ν, over O. Consider its canonical lift to the tangent bundle TO and denote by δ the associated distance on TO. One sees first that, for O small enough, there exists a constant C_1 such that the following regularity estimate holds for Ω on O:

$$|\Omega(u,v) - \Omega(u',v')| \leq C_1\,\delta(u,u')\,\delta(v,v'),$$

for all u, v, u', v' in TO.

Since ν is C^1, one can still assert that the (image of the) stable leaf Δ_a^+ is tangent to the (image of the) fiber E_a^+ of E^+ at the point $a \in V$. One easily shows, using a finite open covering trivializing the tangent bundle over V, that there exists a constant C_2 such that for all $a \in V$, $b \in \Delta_a^+$ and $u \in T_b \Delta_a^+$, one can find a point $b' \in E^+$ and a tangent vector $u' \in T_{b'} E_a^+$ such that

$$\delta(u, u') \leq C_2\,\delta(b, V)\,\|u\|,$$

where $\|u\| = \delta(u, 0_b)$. If $u, v \in T_b \Delta_a^+$, we denote by $u_n = T\phi^n(u)$, $v_n = T\phi^n(v)$ their n-th interates. Using the isotropic character of E_a^+, one has for all n:

$$|\Omega(u_n, v_n)| = |\Omega(u_n, v_n) - \Omega(u_n', v_n')| \leq C_1\,\delta(u_n, u_n')\,\delta(v_n, v_n').$$

Now by the very definition of the stable leaves, if $b_n = \phi^n(b)$ one gets $\delta(b_n, T) \leq C_3 \rho^n$ for a suitable C_3, and thus:

$$|\Omega(u_n, v_n)| \leq C_4 \rho^{2n} \|u_n\|\,\|v_n\|.$$

Finally, a simple continuity argument shows the existence of a neighborhood O' of $E^+ \setminus V$ containing the union $\bigcup_{a \in V} \Delta_a^+$, over which the estimate $\|T\phi\| < 1/\rho$ holds. Since the iterates u_k and v_k stay in O', one gets $\Omega(u, v) = \lim_{n \to \infty} \Omega(u_n, v_n) = 0$, and the stable leaf Δ_a^+ is thus isotropic. $\square$

1.8.2. Isotropic hyperbolic invariant manifolds. The last result crucially uses the fact that the symplectic product of the n-th iterates of two tangent vectors at some point of one leaf tends to zero when $n \to \infty$, and in that case this property comes simply from the convergence to zero of the lengths of these iterates. If one assumes now that the invariant manifold V is *isotropic* one can expect some improvements of the result, since the vanishing of the symplectic product is in that case a direct consequence of the convergence of the iterates to the isotropic tangent bundle TV. More precisely we have:

PROPOSITION. *Let V be a (ρ, l) regularly hyperbolic manifold, with $\rho \in \,]0,1[$ and $l \geq 1$. Assume furthermore that $M(T\phi_{|TV}) < 1/M(T\phi_{|E^+})$, with the usual notation. Then if V is isotropic, the ρ-stable manifold $W^+(V)$ is isotropic too.*

PROOF. Denote by F the tangent bundle of V and by E^+ the stable bundle over V. We have to check first that the vector bundle $G^+ = E^+ \oplus F$ is isotropic. Since F and E^+ are isotropic, it suffices to show that for all $a \in V$, $v \in E_a^+$ and $w \in F_a$, $\Omega(v, w) = 0$. One has $\Omega\big((T\phi)^n(v), (T\phi)^n(w)\big) = \Omega(v, w)$, for all n, just as above. But now we have the inequality:

$$\big|\Omega\big((T\phi)^n(v), (T\phi)^n(w)\big)\big| \leq C\,\|(T\phi)^n(v)\|\,\|(T\phi)^n(w)\|,$$

and our assumptions on the norms ensures that the r.h.s. tends to 0 as $n \to +\infty$. This yields $\Omega(v, w) = 0$ and G^+ is isotropic. One then uses the tangency of $W^+(V)$ to the bundle G^+ to prove the proposition, exactly in the same way as in 1.8.1. $\square$

1.8.3. Normally hyperbolic invariant manifolds.
We now examine the consequences of normal hyperbolicity for an invariant manifold V. The striking fact is that one gets geometric information not only on the stable (or unstable) manifolds of V, but also on the manifold V itself. Namely we have:

PROPOSITION. *Let V be a s-normally hyperbolic invariant submanifold of M, with $s \geq 2$. Then V is symplectic, and the stable and unstable manifolds $W^\pm(V)$ are coisotropic. Furthermore, the leaves of the characteristic foliation of $W^+(V)$ and $W^-(V)$ are exactly the stable and unstable leaves Δ_a^+ and Δ_a^- respectively.*

Let us first say some words about the properties of coisotropic submanifolds of a symplectic manifold. Recall first that a submanifold C of M is coisotropic when for each $x \in C$, the tangent space $T_x C$ contains its symplectic orthogonal in $(T_x M, \Omega_x)$. Thus C carries a canonical distribution, namely the field of subspaces $((T_x C)^\perp)_{x \in C}$, called the *characteristic distribution* of C. The typical example is given by a hypersurface $H = \mathrm{cst}$ in M, which is always coisotropic and whose characteristic distribution is generated by the Hamiltonian vector field X_H; in this example the (one-dimensional) distribution is of course integrable, by the usual Cauchy-Lipschitz theorem. It turns out that in fact the characteristic distribution of a coisotropic submanifold C is *always* integrable: this is indeed a simple application of the Frobenius theorem and of Lie derivatives (see for instance [MDS]), and it makes it possible to define a characteristic *foliation* of C.

PROOF. Note first that due to the Hamiltonian character of the vector field, the dimensions $\dim E^+$ and $\dim E^-$ are *equal*, say $k = \dim E^\pm$ (then $\dim V = 2(\ell - k)$). In order to prove the coisotropy of, say, $W^+ = W^+(V)$, one has first to check the property for the bundle $G^+ = TV + E^+$, and then to use the tangency to get the conclusion. We will verify that for each $a \in V$, the orthogonal $(G_a^+)^\perp$ is exactly E_a^+. It will be enough to prove that $E_a^+ \subset (G_a^+)^\perp$, since then, counting dimensions,

$$\dim(G_a^+)^\perp = 2\ell - \dim(G_a^+) = 2\ell - (2\ell - k) = k = \dim E_a^+,$$

and thus $(G_a^+)^\perp = E_a^+ \subset G_a^+$, meaning that G^+ is indeed a *coisotropic* bundle. In order to prove the inclusion, consider two vectors $v \in E_a^+$ and $w \in G_a^+$; then again:

$$\left|\Omega\big((T\phi)^n(v), (T\phi)^n(w)\big)\right| \leq C \, \|(T\phi)^n(v)\| \, \|(T\phi)^n(w)\| \leq C\rho^n \mu^n \|v\| \, \|v\|$$

for suitable ρ and μ, with $\mu < 1/\rho$. The symplectic product thus tends to zero when $n \to +\infty$, and $\Omega(v, w) = 0$. The end of the proof of the coisotropy of W^+ goes exactly as before.

As for the symplectic character of V, note that the tangent space $T_a V$ is the intersection of the coisotropic spaces $T_a W^\pm(V) = G_a^\pm$; its symplectic orthogonal is thus given by:

$$(T_a V)^\perp = (G_a^+)^\perp + (G_a^-)^\perp = E_a^+ + E_a^-,$$

and one gets $(T_a V)^\perp \cap T_a V = \{0\}$. This shows that the restriction to $T_a V$ of the symplectic form Ω is nondegenerate and V is a $2(\ell - k)$-dimensional *symplectic* submanifold of M. □

1.8.4. Hyperbolic tori.
We now specialize the notion of hyperbolicity to the case of invariant tori and apply the preceding results to that case.

DEFINITION. Let k be an integer, $k \geq 1$. An invariant torus $\mathcal{T} \subset M$ is said to be k-hyperbolic if there exists a continuous decomposition $T_{\mathcal{T}}M = E^+ \oplus E^0 \oplus E^-$ and a Riemannian metric $\| \; \|$ on M such that:
i) $\dim E^+ = \dim E^- = k$;
ii) the tangent bundle of $\mathcal{T}$ is contained in E^0;
iii) $M(T\phi_{|E^+}) < 1$, $m(T\phi_{|E^-}) > 1$;
iv) $M(T\phi_{|E^0}) = m(T\phi_{|E^0}) = 1$.

Let $\mathcal{T}$ be a k-hyperbolic torus, and set

$$\rho_0 = \mathrm{Sup}(M(T\phi_{|E^+}), 1/m(T\phi_{|E^-})),$$

then for all $\rho \in]\rho_0, 1[$, $\mathcal{T}$ is ρ-hyperbolic for ϕ and ϕ^{-1}. We say that a k-hyperbolic torus is l-regular when its invariant manifolds $W^\pm(\mathcal{T})$ are C^l-submanifolds of M. For the sake of simplicity we here confine ourselves to the C^∞-regular case, and simply say regular for C^∞-regular.

Proposition 1.8.1 immediately applies and leads to the

COROLLARY. *Let $\mathcal{T} \subset M$ be a k-hyperbolic invariant torus, with C^1 subbundles E^+ and E^-. Then the stable and unstable leaves Δ_a^+ and Δ_a^- attached to its points are isotropic.*

One would like to apply Proposition 1.8.2, thereby getting some additional information on the geometry of invariant tori. As a preliminary step, we first recall a classical geometric result ([H2], [Be3]) which does not require any hyperbolicity assumption on the invariant object.

PROPOSITION. *Let $\mathcal{T} \subset M$ be a d-dimensional torus which is invariant under Φ, and such that the flow on $\mathcal{T}$ is conjugate to a minimal rotation. If the form Ω is exact over some neighbourhood of $\mathcal{T}$, then $\mathcal{T}$ is isotropic (so that $d \leq \ell$).*

PROOF. There exists an embedding $f : \mathbb{T}^d \hookrightarrow M$ with $f(\mathbb{T}^d) = \mathcal{T}$ and such that $X = f^*(X_H)$ is a (nonresonant) constant vector field. We denote by $\overline{\Omega}$ the pull-back $f^*(\Omega)$. The form $\overline{\Omega}$ is invariant under the flow of X so that the coefficients of $\overline{\Omega}$ in a canonical coordinate system $(x_1, \ldots, x_d)$ of $\mathbb{T}^d$ are constant. One can thus write the form $\overline{\Omega}$ as $\overline{\Omega} = \sum_{1 \leq i < j \leq d} a_{ij} \, dx_i \wedge dx_j$, with $a_{ij} = \pm \int_{\mathbb{T}^d} \overline{\Omega} \wedge \eta_{ij}$, where $\eta_{ij} = dx_1 \wedge \ldots \wedge \widehat{dx_i} \wedge \ldots \wedge \widehat{dx_j} \wedge dx_d$.

Now $\overline{\Omega}$ is an exact form on $\mathbb{T}^d$, because denoting by λ a primitive of Ω near $\mathcal{T}$, we have: $\overline{\Omega} = f^*(d\lambda) = d f^*(\lambda)$. The form $\overline{\Omega} \wedge \eta_{ij}$ is thus exact too: indeed $d(f^*(\lambda) \wedge \eta_{ij}) = \overline{\Omega} \wedge \eta_{ij}$ (η_{ij} is closed). So by Stokes theorem $a_{ij} = 0$ for all i, j. The form $\overline{\Omega}$ is identically zero on $\mathbb{T}^d$, i.e. $\mathcal{T}$ is isotropic. □

In the more general case of an invariant torus $\mathcal{T}$ with *minimal flow* (all the orbits are dense), the question of the isotropic character of $\mathcal{T}$ seems to be open ([H2]). Note that the assumptions of the proposition are always met in the case of hyperbolic tori whose existence is proved by KAM methods, in the neighborhood of Lagrangian tori. Given the above, in the rest of this paragraph, we confine ourselves to the case of isotropic hyperbolic tori. We now state a useful result which is an immediate application of Proposition 1.8.2:

COROLLARY. *Let $\mathcal{T} \subset M$ be an isotropic k-hyperbolic regular invariant torus. Then its invariant manifolds $\mathcal{W}^+$ and $\mathcal{W}^-$ are isotropic.*

Note that if the d-dimensional isotropic torus $\mathcal{T}$ is k-hyperbolic, then the isotropy of $\mathcal{W}^\pm(\mathcal{T})$ forces the inequality $d + k \leq \ell$. In the sequel we will be especially concerned with the case $d + k = \ell$, which can thus be view as "maximal" from the viewpoint of dynamical as well as geometric characteristics. Indeed in that case the invariant manifolds $\mathcal{W}^\pm(\mathcal{T})$ are *Lagrangian*, and the study of their splitting becomes a particular case of the general study of the above paragraphs.

1.9. The perturbative setting

The aim of this paragraph is to introduce a class of Hamiltonian systems on the cotangent bundle $T^*\mathbb{T}^\ell = T^*\mathbb{T}^n \times T^*\mathbb{T}^m$, which generalizes both Hamiltonian (*) and the classical normal forms in the neighborhood of multiple resonances of multiplicity m, which we will meet in Chapter 2.

1.9.1. Somewhat changing notation for convenience, we will consider Hamiltonian systems of the following form:

$$H(\phi, I, q, p) = F(I, q, p) + \mu G(\phi, I, q, p),$$

with $(\phi, I) \in T^*\mathbb{T}^n$ and $(q, p) \in T^*\mathbb{T}^m$; $\mu > 0$ is a small parameter.

The system described by F is independent of ϕ, so that the I-variables are first integrals of X_F. As a consequence, the levels $\{I = I^0\}$ are invariant *coisotropic* submanifolds of $T^*\mathbb{T}^\ell$, diffeomorphic to $\mathbb{T}^n \times T^*\mathbb{T}^m$. We denote by $F_{|I^0}$ the restriction of F to the set $\{I = I^0\}$, that one can consider as a Hamiltonian system on $T^*\mathbb{T}^m$ since it does not depend on ϕ. We will assume that $O = (0, 0) \in T^*\mathbb{T}^m$ is a hyperbolic fixed point for the vector field generated by $F_{|0}$.

As O is hyperbolic, for each I in a neighborhood B of 0 in $\mathbb{R}^n$, we find a family of fixed points $O(I)$ for the systems generated on $T^*\mathbb{T}^m$ by $F_{|I}$; we denote by $W^\pm(O(I))$ their invariant manifolds. This implies that the submanifold:

$$\mathbf{N}_0 = \{(\phi, I, O(I)), \phi \in \mathbb{T}^n, I \in B\} \subset T^*\mathbb{T}^\ell$$

is invariant under the flow defined by F, and one checks that it is a *normally hyperbolic* submanifold of $T^*\mathbb{T}^\ell$, with stable and unstable manifolds:

$$\mathbf{W}_0^\pm = \bigcup_{I \in B} W^\pm(O(I)).$$

One verifies easily in that simple case the assertions of Proposition 1.8.3; namely $\mathbf{N}_0$ is symplectic and in fact symplectomorphic to a neighborhood of the zero section in the cotangent bundle $T^*\mathbb{T}^n$; $\mathbf{W}_0^+$ and $\mathbf{W}_0^-$ are coisotropic, the leaves of the characteristic foliation of $\mathbf{W}_0^\pm$ being the sets:

$$\Delta^\pm_{(\phi^0, I^0, O(I^0))} = \{\phi = \phi^0,\ I = I^0\} \cap \mathbf{W}_0^\pm.$$

One has furthermore in this unperturbed system a foliation of the manifold $\mathbf{N}_0$ by *invariant tori*:

$$\mathcal{T}_{I^0} = \{(\phi, I^0, O(I^0)), \phi \in \mathbb{T}^n\},$$

which are obviously isotropic and m-hyperbolic, with Lagrangian invariant manifolds:

$$\mathcal{W}^\pm(\mathcal{T}_{I^0}) = \{I = I^0\} \cap \mathbf{W}_0^\pm.$$

Let $\mathcal{F}(I^0)$ be the level set of F containing the unperturbed torus $\mathcal{T}_{I^0}$, so $\mathcal{F}(I^0) = F^{(-1)}(\{F(I^0, O(I^0))\})$. The manifold $\mathcal{C}(I^0) = \{I = I^0\} \cap \mathcal{F}(I^0)$ is coisotropic and $(2m + n - 1)$−dimensional; generalizing §1.1.10 one proves easily that the dimension of its characteristic foliation is $2\ell - (2m + n - 1) = n + 1$. So any two Lagrangian submanifolds of $\mathcal{F}(I^0)$ either have empty intersection or intersect along the $(n+1)$−dimensional leaves of the characteristic foliation. In particular, the dimension of the intersection of $\mathcal{W}^+(\mathcal{T}_{I^0})$ and $\mathcal{W}^-(\mathcal{T}_{I^0})$ is at least $n+1$, if there exists a homoclinic point. Due to the very particular form of F, one can easily describe that intersection: it is just the product by $\mathbb{T}^n$ of the homoclinic set $W^+(O(I^0)) \cap W^-(O(I^0))$.

Let us now briefly consider the system decribed by the full perturbed Hamiltonian H. One would like to use Theorem 1.2.3 and Proposition 1.8.3 to deduce, for μ small enough, the existence of a normally hyperbolic submanifold $\mathbf{N}_\mu$, close to $\mathbf{N}_0$. The problem is that $\mathbf{N}_0$ is not compact, so one has to modify slightly the situation. Note first that our study is essentially local, as we are interested only in one single torus contained in the perturbed system. So one has just to modify the system in the neighborhood of the boundary of B in order to get the desired stability, for instance by considering a compact-supported perturbation, equal to the analytic one in a neighborhood of the center of B. After such a modification our theorems apply and we get a normally hyperbolic symplectic submanifold $\mathbf{N}_\mu$, with its invariant stable and unstable coisotropic manifolds $\mathbf{W}_\mu^+$ and $\mathbf{W}_\mu^-$. *In the sequel will always assume that our system has been modified this way, and that we get genuine invariant manifolds in the perturbed system.*

As for the perturbed tori, one cannot of course assert the persistence of the complete family $\mathcal{T}_{I^0}$, but one gets by KAM theorem a large family of invariant *isotropic* tori $\mathcal{T}_{I^0}(\mu)$, located on the invariant manifold $\mathbf{N}_\mu$ and issuing from the unperturbed ones. One easily checks that these are m-hyperbolic tori according to Definition 1.8.4, and one can assert directly in that case that their (Lagrangian) invariant manifolds are μ-close to those of the unperturbed family, since the invariant manifold attached to a given torus is obtained as the collection of the invariant leaves of its points, which are μ-close to the unperturbed ones by the *normal* hyperbolicity of $\mathbf{N}_0$. We will return to this picture in a more analytic and quantitative fashion in Chapter 2.

Let us pass to the assumptions relative to the homoclinic orbits. Here we will adopt a setting which can be seen as an intermediate between purely variational and purely hyperbolic approaches. Generally speaking, it amounts essentially to assuming the existence of enough hyperbolicity and transversality at the averaged level, and then to get the existence of homoclinics (and other objects which we will not consider here) in the initial system by using symplectic geometry. Our assumptions are still "generic", and provide an interesting geometric setting to describe the hyperbolic behaviour of the initial system.

More precisely, in the case of our system $H(I, \phi, p, q)$, we will concentrate on the homoclinic orbits asymptotic to the perturbed torus $\mathcal{T}_0(\mu)$. In order to get the existence of such homoclinics, we will assume, in addition to the hyperbolicity of the fixed point $O = (0,0)$ in $T^*\mathbb{T}^m$, the existence of a homoclinic orbit ξ biasymptotic to O (for the unperturbed system $X_{F_{|0}}$ on $T^*\mathbb{T}^m$), such that
– the stable and unstable manifolds $W^\pm(O)$ intersect transversely along ξ;
– the manifolds $W^\pm(O)$ are *graphs* in the neighborhood of ξ.

Under these assumptions, we will prove in §1.10 the existence of homoclinic orbits for the perturbed torus, using a suitable section and some basic facts of symplectic geometry; and we will examine in §1.11 various ways for defining the splitting. We end the present paragraph on the perturbative setting by some straightening results for the various invariant manifolds we just described; these results will prove useful for the application of KAM theorems in Chapter 2.

1.9.2. Straightening theorems in the perturbative setting. Here we show how to simultaneously straighten the stable and unstable invariant manifolds, in the perturbative setting and under various circumstances. We sketch and emphasize the rather uniform procedures, refraining from giving formal statements which might be longer and more complicated than the proofs. These results will prove useful in particular in Chapter 2.

As a first (nonperturbative!) example, we begin with the case of a *hyperbolic fixed point* $O \in M$, for which we want to locally straighten the stable and unstable manifolds simultaneously. Proposition 1.8.1 shows that these invariant manifolds $\mathcal{W}^\pm$ are isotropic, thus Lagrangian since they are ℓ-dimensional; this is also the case for the stable and unstable directions $E^\pm$ of the linearized system, in the symplectic space $(T_O M, \Omega_O)$. Since this is a local study, we may assume that $M = \mathbb{R}^{2\ell}$ (with the standard form Ω) and $O = 0$, and due to the Lagrangian character of $E^\pm$ we may assume furthermore that $E^+ = \mathbb{R}^\ell \times \{0\}$ and $E^- = \{0\} \times \mathbb{R}^\ell$. Now denoting by $(\bar{s}, \bar{u})$ the variables in $\mathbb{R}^{2\ell}$, one can represent the *local* invariant manifolds $\mathcal{W}_l^\pm$ as *graphs* over the stable and unstable directions:

$$\mathcal{W}_l^+ = \{(\bar{s}, g_+(\bar{s})),\ \bar{s} \in D\}, \qquad \mathcal{W}_l^- = \{(g_-(\bar{u}), \bar{u}),\ \bar{u} \in D\},$$

where D is some open disk in $\mathbb{R}^\ell$, and $Dg_\pm(0) = 0$. The function $j_+ : D \to \mathbb{R}^{2\ell}$ defined by $j_+(\bar{s}) = (\bar{s}, g_+(\bar{s}))$ is thus a *Lagrangian* embedding of D, as well as the function j_- defined by $j_-(\bar{u}) = (g_-(\bar{u}), \bar{u})$, so that:

$$j_+^* \Omega = j_-^* \Omega = 0.$$

We now want to straighten $\mathcal{W}^+$ and $\mathcal{W}^-$, *i.e.* to introduce symplectic coordinates (s, u) in a neighborhood of 0 in which the local equations simply read:

$$\mathcal{W}_l^+ = \{u = 0\}, \qquad \mathcal{W}_l^- = \{s = 0\}.$$

We will obtain these coordinates by composing two natural symplectic diffeomorphisms ρ_+ and ρ_-, straightening first the stable manifold *via* ρ_+ without loosing the graph property for $\mathcal{W}^-$, then straightening $\mathcal{W}^-$ *via* ρ_- without perturbing $\mathcal{W}^+$.

The map $(\bar{s}, \bar{u}) \mapsto (\bar{s}, \bar{u} - g_+(\bar{s}))$ defines a diffeomorphism ρ_+ in a neighborhood of 0, since its derivative at 0 is the identity map, and one obviously gets for $\rho_+(\mathcal{W}^+)$ the equation $\bar{u} = 0$. One checks furthermore by a direct calculation that if π is the projection $(\bar{s}, \bar{u}) \mapsto \bar{s}$:

$$\rho_+^* \Omega = \Omega - (j_+ \circ \pi)^* \Omega = \Omega - \pi^* j_+^* \Omega = \Omega,$$

and ρ_+ is a *symplectomorphism*. Note that here of course the Lagrangian character of the stable manifold was used in a crucial way. One can thus introduce *via* ρ_+ a symplectic coordinate system $(\tilde{s}, \tilde{u})$ in a neighborhood of 0 in which $\mathcal{W}^+$ has the trivial equation $\tilde{u} = 0$: this takes care of the straightening of the stable manifold.

In these new coordinates $(\tilde{s}, \tilde{u})$, the manifold $\mathcal{W}^-$ is still a graph over the vertical space, since its equation $\bar{s} = g_-(\bar{u})$ transforms into $\tilde{s} = g_-(\tilde{u} + g_+(\tilde{s}))$, which by the Implicit Function Theorem is locally equivalent to $\tilde{s} = \overline{g}_+(\tilde{u})$ for a suitable function $\overline{g}_-$. Here we have used the equality $Dg_+(0) = 0$, reflecting the tangency of the stable manifold to the stable direction.

One can now apply the same process to the unstable manifold in the coordinates $(\tilde{s}, \tilde{u})$, and introduce the new symplectic chart $(s = \tilde{s} - \overline{g}_-(\tilde{u}), u = \tilde{u})$ in a neighborhood of 0. One directly checks that both stable and unstable manifolds take the announced form in the coordinates (u, s). Note finally that these coordinates have the same regularity as the invariant manifolds; in particular they are *analytic* if the manifolds are.

We now take up the case of a *hyperbolic isotropic torus* in the perturbative setting described in §1.9.1. Given a m-hyperbolic torus $\mathcal{T}$, one can first find a symplectic coordinate system $(\phi, I, \bar{s}, \bar{u})$, with $(\phi, I) \in T^*\mathbb{T}^n$ and $(\bar{s}, \bar{u}) \in \mathbb{R}^{2m}$, $n + m = \ell$, such that the torus $\mathcal{T}$ and its stable and unstable manifolds $\mathcal{W}^{\pm}$ appear as graphs of suitable functions:

$$\mathcal{T} = \{(\phi, I(\phi), S(\phi), U(\phi)), \ \phi \in \mathbb{T}^n\},$$
$$\mathcal{W}^+ = \{(\phi, I^+(\phi, \bar{s}), \bar{s}, U^+(\phi, \bar{s})), \ \phi \in \mathbb{T}^n, \ \bar{s} \in D\},$$
$$\mathcal{W}^- = \{(\phi, I^-(\phi, \bar{u}), S^-(\phi, \bar{u}), \bar{u},), \ \phi \in \mathbb{T}^n, \ \bar{u} \in D\}.$$

Exactly as in the case of a fixed point, one can perform two transformations in order to straighten $\mathcal{W}^+$ and $\mathcal{W}^-$ successively, together with also their intersection $\mathcal{T}$. As for the stable manifold, one introduces the local diffeomorphism $\rho_+ : (\phi, I, \bar{s}, \bar{u}) \mapsto (\tilde{\phi}, \tilde{I}, \tilde{s}, \tilde{u})$ defined by:

$$\tilde{\phi} = \phi, \ \tilde{I} = I - I^+(\phi, \bar{s}), \ \tilde{s} = \bar{s}, \ \tilde{u} = \bar{u} - U^+(\phi, \bar{s}),$$

in a neighborhood of $\mathcal{T}$. In the new coordinates $(\tilde{\phi}, \tilde{I}, \tilde{s}, \tilde{u})$ the manifold $\mathcal{W}^+$ is clearly straightened, with equation $\tilde{I} = 0$, $\tilde{u} = 0$. Now the Lagrangian character of $\mathcal{W}^+$ implies again that ρ_+ is symplectic: more precisely, let π now denote the projection $(\phi, I, \bar{s}, \bar{u}) \mapsto (\phi, 0, \bar{s}, 0)$ and j_+ the Lagrangian embedding:

$$(\phi, \bar{s}) \mapsto (\phi, I^+(\phi, \bar{s}), \bar{s}, U^+(\phi, \bar{s})).$$

Then one has, just as above in the case of the fixed point:

$$\rho_+^* \Omega = \Omega - (j_+ \circ \pi)^* \Omega = \Omega.$$

As before one again checks that $\mathcal{W}^-$ is still a graph on the $(\phi, \bar{u})$ space, so that the same idea may be applied again and leads to a new symplectic system (ψ, J, s, u) in the neighborhood of $\mathcal{T}$, in which the local equations for the invariant manifolds are trivial:

$$\mathcal{W}^+ = \{J = 0, \ u = 0\}, \quad \mathcal{W}^- = \{J = 0, \ s = 0\}.$$

This in turn leads to the straightening of $\mathcal{T}$, because $\mathcal{T} = \mathcal{W}^+ \cap \mathcal{W}^- = \{J = 0, \ s = u = 0\}$. Finally the coordinates here are again as regular as the invariant manifolds themselves.

Next it is natural to consider *normally hyperbolic manifolds*, still in the perturbative setting, namely the unperturbed and perturbed invariant manifolds $\mathbf{N}_0$ and $\mathbf{N}_\mu$ introduced in §1.9.1. Due to the invariance of the Hamiltonian F relative to the variable ϕ, the first case may be treated almost exactly as above, while the second one is more complicated and will not be considered here.

We start with the straightening of the *unperturbed normally hyperbolic manifold* $\mathbf{N}_0$ and its attending stable and unstable invariant manifolds $\mathbf{W}_0^\pm$. We work on a domain over which the manifolds $\mathbf{W}_0^\pm$ are graphs over their tangent space along $\mathbf{N}_0$, so that in a suitable system of coordinates $(\phi, I, \bar{s}, \bar{u})$ we can represent $\mathbf{W}_0^+$ as:

$$\mathbf{W}_0^+ = \{(\phi, I, \bar{s}, U(I, \bar{s})),\ \phi \in \mathbb{T}^n,\ I \in D^n,\ s \in D^m\}$$

where D^n and D^m are some open sets in $\mathbb{R}^n$ and $\mathbb{R}^m$. Here we have used the independence of the (unperturbed) system on the angle coordinates ϕ. We introduce, much as above, the local diffeomorphism $\rho_+ : (\phi, I, \bar{s}, \bar{u}) \mapsto (\tilde{\phi}, \tilde{I}, \tilde{s}, \tilde{u})$ defined by

$$\tilde{\phi} = \phi,\ \tilde{I} = I,\ \tilde{s} = \bar{s},\ \tilde{u} = \bar{u} - U(I, \bar{s}).$$

The pull-back of the form Ω by ρ_+ now reads

$$\rho_+^* \Omega = dI \wedge d\phi + d\bar{u} \wedge d\bar{s} - dU \wedge d\bar{s},$$

and one directly checks that the last term vanishes, since the characteristic foliation of $\mathbf{W}_0^+$ is precisely parameterized by the variable $\bar{s}$. Thus ρ_+ is symplectic again and it obviously straightens $\mathbf{W}_0^+$. One can then repeat the process as before and introduce a symplectic coordinate system (ψ, J, u, s) in which the local equations are again trivial:

$$\mathbf{W}_0^+ = \{u = 0\}, \qquad \mathbf{W}_0^- = \{s = 0\},$$

and consequently

$$\mathbf{N}_0 = \{u = s = 0\}.$$

In this example the regularity of the coordinates is still the same as the regularity of $\mathbf{W}_0^\pm$.

1.10. Lagrangian intersections and homoclinic trajectories

We address in this paragraph the question of the existence of homoclinic orbits for hyperbolic tori and we wish to simultaneously cover the "non-integrable" cases, where the invariant manifolds of the tori are distinct, as well as the "integrable" ones, where they may coincide. As a consequence we cannot make use of the classical methods because these are based on a transversality assumption for the intersections. We will instead use in a crucial way the fact that the invariant manifolds of isotropic tori are Lagrangian and translate the problem of existence of homoclinic trajectories into a problem of Lagrangian intersections. In other words we do *not* use general hyperbolic methods but really exploit the symplectic features of the problem. In the case of 1-hyperbolic KAM tori (*i.e.* in the nondegenerate perturbative setting), a complete proof was given by H.Eliasson in [El] (see also [DG1]). We give here a simple proof valid for all multiplicities, which is not a "complete" proof since we will make use (when $m \geq 2$) of a classical variational result ensuring the existence of homoclinic orbits relative to the hyperbolic fixed point in a multidimensional pendulum (see below). For $m \geq 2$ we find this setting most interesting since, under the generic assumption of transversality along the homoclinics in the pendulum, one can prove the existence of a very rich hyperbolic behaviour, in the pendulum first (see [Mar2] and [ST]), and then in the initial system. This study should be continued in connection with the problem of exchange of resonances, along the lines of [Mar2].

1.10.1. Homoclinic orbits to 1-hyperbolic tori in the perturbative setting. In this subsection and the next one we will use the perturbative setting described in §1.9.1. On the cotangent bundle $T^*\mathbb{T}^\ell = T^*\mathbb{T}^n \times T^*\mathbb{T}^m$ we consider the Hamiltonian:

$$H(\phi, I, q, p) = F(I, q, p) + \mu G(\phi, I, q, p)$$

and denote by $\hat{F}$ the restriction of F to the set $\{I = 0\}$, considered as a Hamiltonian system on $T^*\mathbb{T}^m$. In this paragraph we study the case $m = 1$ which has specific features which simplify both the description of the system and the proof of the existence of the homoclinic orbits.

When $m = 1$, the system $\hat{F}$ is defined on the cotangent bundle $T^*\mathbb{T}$, and thus completely integrable. We emphasize that we do not actually make use of such a regular structure and the following assumptions in fact suffice for our purposes:
– the point $O = (0, 0)$ is a hyperbolic fixed point with $\hat{F}(O) = 0$;
– there exists a homoclinic orbit ξ biasymptotic to O;
– the set $\xi \cup \{0\}$ is the graph of a function $X : \mathbb{T}^1 \to \mathbb{R}$.

The Hamiltonian vector field does not vanish on the orbit ξ, so that we may assume that its projection on $\mathbb{T}^1$ is compatible with the natural orientation. We fix on the torus (circle) $\mathbb{T}^1$ two open neighborhoods D^+ and D^- of $[\pi, 2\pi = 0]$ and $[0, \pi]$ respectively; we think of D^+ (resp. D^-) as the arc along which we will follow the stable manifold (resp. the unstable one) in the perturbed system.

For the complete system F on $T^*\mathbb{T}^\ell$, the above assumptions lead naturally to the existence of corresponding invariant objects, obtained from the previous ones by direct product with the torus $\mathbb{T}^n$. One gets an invariant 1-hyperbolic torus:

$$\mathcal{T}_0 = \{(\phi, 0, 0, 0), \ \phi \in \mathbb{T}^n\},$$

whose stable and unstable manifolds *coincide*:

$$\mathcal{W}_0^+ = \mathcal{W}_0^- = \{(\phi, 0, q, X(q)), \ \phi \in \mathbb{T}^n, \ q \in \mathbb{T}^1\}.$$

Now, for μ small enough, we *assume* the existence in the perturbed system, as described by the function H, of a hyperbolic torus $\mathcal{T}_\mu$ close to $\mathcal{T}_0$, with stable and unstable manifolds $\mathcal{W}_\mu^\pm$ close to $\mathcal{W}_0^\pm$, and we will prove that they intersect:

PROPOSITION. *There exists $\mu_0 > 0$ such that for all $\mu \in {]}0, \mu_0{[}$, the torus $\mathcal{T}_\mu$ has at least ℓ homoclinic orbits.*

PROOF. Let us make our closeness assumption and our notation more precise. In the rest of this paragraph the parameter μ will be fixed (small enough), and we abbreviate $\mathcal{T}_\mu$ and $\mathcal{W}_\mu^\pm$ into $\mathcal{T}$ and $\mathcal{W}^\pm$. Now the torus and its manifolds are close to the unperturbed ones, and one gets three functions $I : \mathbb{T}^n \to \mathbb{R}^n$, $Q : \mathbb{T}^n \to \mathbb{R}$, $P : \mathbb{T}^n \to \mathbb{R}$ such that:

$$\mathcal{T} = \Big\{ \big(\phi, I(\phi), Q(\phi), P(\phi)\big), \ \phi \in \mathbb{T}^n \Big\},$$

and (twice) two functions $I^\pm : \mathbb{T}^n \times D^\pm \to \mathbb{R}^n$ and $P^\pm : \mathbb{T}^n \times D^\pm \to \mathbb{R}$ such that:

$$\mathcal{W}^\pm = \Big\{ \big(\phi, I^\pm(\phi, q), q, P^\pm(\phi, q)\big), \ \phi \in \mathbb{T}^n, \ q \in D^\pm \Big\}.$$

The main idea will be to construct a section which is well adapted to the problem. Remark first that the nonvanishing of the Hamiltonian vector field on the homoclinic orbit ensures that $\partial F/\partial p(0, \pi, X(\pi)) \neq 0$, and the Implicit Function

Theorem leads to the existence of a local parametrization $p = f(I, q)$ of the level surface $F^{-1}(0)$ in the neighborhood of $\mathbb{T}^n \times \{(0, \pi, X(\pi))\}$. Considering now the hypersurface S with equation $q = \pi$ in $T^*\mathbb{T}^\ell$, one gets a local parametrization of the intersection $\mathcal{S}_0 = S \cap F^{-1}(0)$, of the form $p = g(I)$, with I in an open neighborhood B_0 of 0 in $\mathbb{R}^n$. One can thus choose (ϕ, I) as coordinates on $\mathcal{S}_0$, and one sees immediately that in these coordinates the one-form induced on $\mathcal{S}_0$ by the Liouville form λ of $T^*\mathbb{T}^\ell$ is the canonical form $I\,d\phi$. One can thus identify $\mathcal{S}_0$ with an open set of the cotangent bundle $T^*\mathbb{T}^n$ with its Liouville structure λ_n. Now for μ small enough the perturbed energy level $H^{-1}(0)$ still intersects the hypersurface S transversely, and if $\mathcal{S} = S \cap H^{-1}(0)$, one sees that one can still choose the variables (ϕ, I) as *exact symplectic* coordinates on $\mathcal{S}$, identifying $\mathcal{S}$ with a neighborhood of the zero section in $(T^*(\mathbb{T}^n), \lambda_n)$. In these coordinates, the intersections of the perturbed manifolds $\mathcal{W}^\pm$ with $\mathcal{S}$ are easy to describe:

$$w^\pm = \left\{ (\phi, \bar{I}^\pm(\phi)), \ \phi \in \mathbb{T}^n \right\}.$$

where $\bar{I}^\pm(\phi) = I^\pm(\phi, \pi)$. The manifolds $\mathcal{W}^\pm$ are Lagrangian, and so their intersections w^+ and w^- with $\mathcal{S}$ are Lagrangian *for the induced structure*, thus Lagrangian in $T^*\mathbb{T}^n$ with the above identification. In order to prove an intersection property, we analyze their defects of exactness $\delta^\pm$ (see §1.7.1), and more precisely we prove that $\delta^+ = \delta^- = \delta$.

To see this we first recall from §1.7.1 that the i-th component $\delta_i^\pm$ is the period of the Liouville form λ_n over the cycle

$$C_i^\pm = \{(\phi, \bar{I}^\pm(\phi)), \ \phi \in \Gamma_i\} \subset w^\pm,$$

where Γ_i is the fundamental cycle $\{\phi_j = 0, \ j \neq i\}$ on $\mathbb{T}^n$. Since λ_n is induced by the ambient form λ on $\mathcal{S}$, $\delta_i^\pm$ is also the period of λ on the same cycle, considered as a subset of $T^*\mathbb{T}^\ell$. The equalities $\delta^+ = \delta^- = \delta$ will now follow from the existence of a homotopy h^+ (resp. h^-) between w^+ and $\mathcal{T}$ (resp. w^- and $\mathcal{T}$) with values in $\mathcal{W}^+$ (resp. $\mathcal{W}^-$), which transforms the cycle C_i^+ (resp. C_i^-) into the cycle (in the $+$ as well as in the $-$ case)

$$C_i = \{(\phi, I(\phi), Q(\phi), P(\phi)), \ \phi \in \Gamma_i\} \subset \mathcal{T}$$

(returning to the orginal coordinates). Namely, in the case, say, of $\mathcal{W}^-$, let:

$$h^- : [0, 1] \times \mathbb{T}^d \to \mathcal{W}^-, \quad h^-(s, \phi) = \Big(\phi, I^-\big(\phi, q(s, \phi)\big), q(s, \phi), P^-\big(\phi, q(s, \phi)\big)\Big),$$

with

$$q(s, \phi) = (1 - s)\,Q(\phi) + s\,\pi.$$

One checks that h^- transforms the cycle C_i^- into C_i and takes its values in $\mathcal{W}^-$. The stable counterpart is constructed analogously. Since the two-form Ω vanishes on $\mathcal{W}^\pm$, one gets by Stoke's theorem that δ_i^+ and δ_i^- are both equal to the period of λ over the cycle C_i, and this proves the equality of the defects of exactness: $\delta^+ = \delta^- = \delta$.

The conclusion easily follows: the manifolds $\overline{w}^\pm = w^\pm - \delta$ are now *exact* Lagrangian submanifolds of $T^*\mathbb{T}^n$, so they have generating functions $S^\pm$, and the critical points of the difference $S^+ - S^-$ are in one-to-one correspondence with the intersection points of $\overline{w}^\pm$, thus also with the ones of $w^\pm$. It is well-kwown (by

Lyusternik-Schnirelman category theory) that a smooth function on the torus $\mathbb{T}^n$ has at least $n+1$ critical points, which proves our proposition since $\ell = n+1$. □

1.10.2. Homoclinic orbits to m-hyperbolic tori in the perturbative setting. We now generalize the above approach to the case $m > 1$. The method is essentially the same: the homoclinic points will be found as critical points of the difference of two generating functions associated with the stable and unstable manifold in a suitable section contained in the energy level of the torus. The main difference is that now the unperturbed invariant manifolds $\mathcal{W}_0^\pm$ of the n-dimensional torus $\mathcal{T}_0$ do not coincide anymore, but instead intersect along an $(n+1)$-dimensional homoclinic manifold ($n+m=\ell$). One now has to find a $2n$-dimensional section Σ, symplectomorphic to $T^*\mathbb{T}^n$, in such a way that the intersection $\Sigma \cap \mathcal{W}^+ \cap \mathcal{W}^-$ coincides with the zero section of $T^*\mathbb{T}^n$. Then the same deformation argument as before will prove that the defects of exactness of the intersections $\mathcal{W}^+ \cap \Sigma$ and $\mathcal{W}^- \cap \Sigma$ coincide, and thus that $\mathcal{W}^+ \cap \mathcal{W}^- (\cap \Sigma)$ is nonempty.

Let us again make clear what the hypotheses are, with the same notation as above. We assume that:
– the point $O = (0,0)$ is a hyperbolic fixed point with $\hat{F}(O) = 0$;
– there exists a homoclinic orbit ξ biasymptotic to O;
– the stable and unstable manifolds $W^\pm$ attached to O intersect transversely along ξ;
– the manifolds $W^\pm$ are *graphs* in the neighborhood of ξ.

In order to make this last assumption more precise, we will parameterize the projection $\gamma = \pi \circ \xi$ of the homoclinic solution on $\mathbb{T}^m$, so as to distinguish between its "stable" part and its "unstable "one. The image $\Gamma = \gamma(\mathbb{R})$ is a curve in $\mathbb{T}^m$, naturally oriented by the projection of the Hamiltonian vector field. We fix some point a on γ and define the positive and negative parts Γ^+, Γ^- of Γ as the set of points located after and before a on Γ, respectively. We introduce then two contractible open neighborhoods D^+ and D^- of Γ^+ and Γ^- in the torus $\mathbb{T}^m$, over which the manifolds $W^\pm$ are graphs over the base $\mathbb{T}^m$, viz.:

$$W^\pm = \{(q, P_0^\pm(q)), \ q \in \mathbb{T}^m\},$$

for two functions $P_0^\pm : D^\pm \to \mathbb{R}^m$. Of course these manifolds are Lagrangian, and even *exact* since the domains of definition of the functions $P_0^\pm$ are contractible. Lastly we denote by $(a, b = P_0^+(a) = P_0^-(a))$ the lift of a to the homoclinic orbit ξ.

For the complete system F on $T^*\mathbb{T}^\ell$ the above assumptions again lead naturally to the corresponding invariant objects. One gets again an invariant n-dimensional and m-hyperbolic torus

$$\mathcal{T}_0 = \{(\phi, 0, 0, 0), \ \phi \in \mathbb{T}^n\},$$

with invariant stable and unstable Lagrangian manifolds

$$\mathcal{W}_0^\pm = \{(\phi, 0, q, P_0^\pm(q)), \ \phi \in \mathbb{T}^n, q \in D^\pm\},$$

which intersect along the $(n+1)$-dimensional homoclinic manifold $\mathcal{H}_0 = \mathbb{T}^n \times \{0\} \times \xi$. Now, for μ small enough and fixed, we assume (again as in §1.10.1) the existence in the *perturbed* system governed by H of a hyperbolic torus $\mathcal{T}_\mu$ close to $\mathcal{T}_0$, with stable and unstable manifolds $\mathcal{W}_\mu^\pm$ close to $\mathcal{W}_0^\pm$; we write again, dropping the subscript μ

as in §1.10.1:
$$\mathcal{T} = \left\{ (\phi, I(\phi), Q(\phi), P(\phi)), \ \phi \in \mathbb{T}^n \right\},$$
with $I : \mathbb{T}^n \to \mathbb{R}^n$, $Q : \mathbb{T}^n \to \mathbb{R}^m$, $P : \mathbb{T}^n \to \mathbb{R}^m$, and
$$\mathcal{W}^\pm = \left\{ (\phi, I^\pm(\phi,q), q, P^\pm(\phi,q)), \ \phi \in \mathbb{T}^n, \ q \in D^\pm \right\},$$
with $I^\pm : \mathbb{T}^n \times D^\pm \to \mathbb{R}^n$ and $P^\pm : \mathbb{T}^n \times D^\pm \to \mathbb{R}^m$. We can now state our result, which exactly parallels and extends Proposition 1.10.1 to any multiplicity m:

PROPOSITION. *There exists $\mu_0 > 0$ such that for all $\mu \in]0, \mu_0[$, the torus $\mathcal{T} = \mathcal{T}_\mu$ has at least $n+1$ homoclinic orbits in the neighborhood of $\mathcal{H}_0$.*

PROOF. The problem consists again in finding a suitable section. We begin at the unperturbed level by picking a hypersurface $\sigma \subset \mathbb{T}^m$ transverse to Γ at the point a; namely $\sigma = \{q_m = a_m\} \cap (D^+ \cap D^-)$ (possibly after permuting the coordinates). The transversality condition implies that $\partial F/\partial p_m(0, a, b) \neq 0$ and the Implicit Function Theorem leads to the existence of a local parametrization $p_m = f(I, q, \bar{p})$ of the level set $F^{-1}(0)$ in the neighborhood of $\mathbb{T}^n \times \{(0, a, b)\}$, with $\bar{p} = (p_1, \ldots, p_{m-1})$. Denoting by S the lift of σ to the total space $T^*\mathbb{T}^\ell$, one gets a local parametrization of the intersection $\mathcal{S}_0 = S \cap F^{-1}(0)$, of the form $p_m = g(I, \bar{q}, \bar{p})$, with I in an open neighborhood B_0 of 0 in $\mathbb{R}^n$, $\bar{q} = (q_1, \ldots, q_{m-1})$ in an open neighborhood $B_{\bar{a}}$ of $\bar{a}$ in $\mathbb{T}^{m-1}$ (or equivalently in $\mathbb{R}^{m-1}$), and $\bar{p}$ in an open neighborhood $B_{\bar{b}}$ of $\bar{b}$ in $\mathbb{R}^{m-1}$. One can thus use $(\phi, I, \bar{q}, \bar{p})$ as coordinates on $\mathcal{S}_0$, and it is plain that in these coordinates the one-form $\hat{\lambda}$ induced on $\mathcal{S}_0$ by the Liouville form λ of $T^*\mathbb{T}^\ell$ reads $\hat{\lambda} = I\,d\phi + \bar{p}\,d\bar{q}$. One can thus identify $\mathcal{S}_0$ with an open set of the cotangent bundle $T^*(\mathbb{T}^n \times \mathbb{R}^{m-1})$ equipped with its Liouville structure.

Now for μ small enough the perturbed energy level $H^{-1}(0)$ still intersects the hypersurface S transversely, and writing $\mathcal{S} = S \cap H^{-1}(0)$, one sees that one can still use the variables $(\phi, I, \bar{q}, \bar{p})$ as *exact* symplectic coordinates on $\mathcal{S}$. In these coordinates the intersections of the perturbed manifolds $\mathcal{W}^\pm$ with $\mathcal{S}$ read:
$$\mathcal{W}^\pm \cap \mathcal{S} = \left\{ (\phi, \bar{I}^\pm(\phi, \bar{q}), \bar{q}, \bar{P}^\pm(\phi, \bar{q})), \ \phi \in \mathbb{T}^n, \ \bar{q} \in \bar{D}^\pm \right\},$$
where $\bar{D}^\pm = \sigma \cap D^\pm$ and $\bar{I}^\pm(\phi, \bar{q}) = I^\pm(\phi, (\bar{q}, a_m))$, $\bar{P}^\pm(\phi, \bar{q}) = P^\pm(\phi, (\bar{q}, a_m))$.

We wish to understand the set $\mathcal{W}^+ \cap \mathcal{W}^- \cap \mathcal{S}$, whose equations in the above coordinates are:

$$(\mathrm{E}_1): \quad \bar{I}^+(\phi, \bar{q}) = \bar{I}^-(\phi, \bar{q}); \qquad (\mathrm{E}_2): \quad \bar{P}^+(\phi, \bar{q}) = \bar{P}^-(\phi, \bar{q}).$$

To this end we first analyze the second equality and construct a new section in which the first equation can be solved using the by now usual Lagrangian intersections argument. Here we make use of the transversality of $W^\pm$ along the homoclinic trajectory ξ and note that for μ small enough Equation (E_2) determines an implicit function $\bar{q} = \bar{Q}(\phi)$, defined on the torus $\mathbb{T}^n$ with values in $\mathbb{R}^{m-1}$, whose graph is the set of solutions of (E_2). Starting from this graph we build our new section, namely an annulus $\Sigma \subset \mathcal{S}$, defined as the image of the following embedding:

$$i : O \subset T^*\mathbb{T}^n \to \mathcal{S}, \qquad (\psi, J) \mapsto \left(\psi, \ J - {}^t D_\psi \bar{Q} \cdot \bar{P}(\psi, \bar{Q}(\psi)), \ \bar{Q}(\psi), \ \bar{P}(\psi, \bar{Q}(\psi)) \right),$$

where O is a suitable open neighborhood of the zero section in $T^*\mathbb{T}^n$, and where the r.h.s. is written using the above coordinates on $\mathcal{S}$. An immediate check confirms that the pull-back $i^*\lambda_{\mathcal{S}}$ of the Liouville form on $\mathcal{S}$ coincides with the Liouville form $\lambda' = J\,d\psi$ on $T^*\mathbb{T}^n$, and we may thus identify the annulus Σ with the neighborhood O.

In the new coordinates (J, ψ), the intersections $w^\pm = \Sigma \cap \mathcal{W}^\pm$ are given by:

$$w^\pm = \Big\{ \big(\psi, \bar{I}^\pm(\psi, \bar{q}) + R(\psi)\big),\ \psi \in \mathbb{T}^n \Big\},$$

where $R(\psi) = {}^t D_\psi \bar{Q} \cdot \bar{P}\big(\psi, \bar{Q}(\psi)\big)$. These being Lagrangian submanifolds of Σ it will suffice to prove that their defects of exactness $\delta^\pm$ coincide, and this will again (compare §1.10.1) be achieved via the use of suitable homotopies with values in the Lagrangian manifolds $\mathcal{W}^\pm$. Let us consider the stable manifold for definiteness. As the expression of $\mathcal{W}^+$ is known only over the domain D^+, we will obtain our homotopy as the concatenation of three homotopies h_i, $i = 1, 2, 3$: the second connects suitable intermediate tori $\mathcal{T}_0^+$ and $\mathcal{T}_1^+$, which are close to $\mathcal{T}$ and w^+ respectively; the first one is between $\mathcal{T}$ and $\mathcal{T}_0^+$, and the last one between $\mathcal{T}_1^+$ and w^+.

We denote by $\mathcal{G}^+$ the natural embedding of the stable manifold $\mathcal{W}^+$, namely the map from $\mathbb{T}^n \times D^+$ to $T^*\mathbb{T}^\ell$ defined by:

$$\mathcal{G}^+(\phi, q) = \big(\phi, I^+(\phi, q), q, P^+(\phi, q)\big).$$

In order to define the homotopy h_2, we introduce the map:

$$\eta_2 : [0,1] \times \mathbb{T}^n \to \mathbb{T}^n \times D^+, \qquad (s, \phi) \mapsto (\phi, \gamma^+(-\log s))$$

and define $h_2 = \mathcal{G}^+ \circ \eta_2$. The map h_2 deforms the torus $\mathcal{T}_0^+ = \{\mathcal{G}^+(\phi, 0),\ \phi \in \mathbb{T}^n\}$ into the torus $\mathcal{T}_1^+ = \{\mathcal{G}^+(\phi, a),\ \phi \in \mathbb{T}^n\}$, with values in $\mathcal{W}^+$ (recall that a denotes some point of γ which was fixed from the start; here we have supposed $\gamma(0) = a$). We now define the following maps:

$$\eta_1 : [0,1] \times \mathbb{T}^n \to \mathbb{T}^n \times D^+, \qquad (s, \phi) \mapsto (\phi, (1-s)\, Q(\phi)),$$

$$\eta_3 : [0,1] \times \mathbb{T}^n \to \mathbb{T}^n \times D^+, \qquad (s, \phi) \mapsto (\phi, (s\,\bar{Q}(\phi) + (1-s)\,\bar{a}, a_m)),$$

and the homotopies $h_1 = \mathcal{G}^+ \circ \eta_1$ and $h_3 = \mathcal{G}^+ \circ \eta_3$. The first one ($h_1$) deforms the torus $\mathcal{T}$ into $\mathcal{T}_0^+$, whereas the last one (h_3) deforms $\mathcal{T}_1^+$ into w^+, both with values in $\mathcal{W}^+$. The concatenation of $h = h_1$ followed by h_2 and then h_3 thus provides the sought-after homotopy between $\mathcal{T}$ and w^+, with values in $\mathcal{W}^+$.

Lastly, we introduce (again as in §1.10.1 above) the cycles C_i and C_i^+, $i = 1, \ldots, n$, defined by

$$C_i = h(0, \Gamma_i) \subset \mathcal{T}, \qquad C_i^+ = h(1, \Gamma_i) \subset w^+,$$

where $\Gamma_i = \{\phi_j = 0,\ j \neq i\}$ is a fundamental cycle of the torus $\mathbb{T}^n$. Since the form $i^*\lambda$ induced by the ambient Liouville form on the annulus Σ is given by $J\,d\psi$, the component δ_i^+ of the defect of exactness of w^+ is given by the period of the Liouville form λ over C_i^+. Stokes theorem now yields the equality

$$\delta_i^+ = \int_{C_i} \lambda.$$

A completely parallel construction along $\mathcal{W}^-$ then shows that $\delta_i^+ = \delta_i^-$ ($= \delta_i$, if this denotes the r.h.s. of the above equality). The defects of exactness of w^+ and

w^- coincide, and the conclusion follows as in the case $m = 1$ of §1.10.1, since the category of $\mathbb{T}^n$ is $n + 1$. □

1.11. The splitting of the invariant manifolds of hyperbolic tori

1.11.1. We specialize now the previous results and describe, in various situations, the splitting of the stable and unstable manifolds attached to a hyperbolic torus. We have proved in §1.8 that the invariant manifolds $\mathcal{W}^{\pm}$ attached to an isotropic hyperbolic torus $\mathcal{T}$ are injectively immersed isotropic submanifolds, hence they are Lagrangian in the "maximally hyperbolic case", to which we will limit ourselves in the sequel. It is also easy to see that when the torus possesses a homoclinic orbit, the pair $(\mathcal{W}^+, \mathcal{W}^-)$ is regular. So the general definitions of §1.6 apply, and we get along the set of homoclinic orbits a symplectic angular form and a Euclidean angular form. Yet a more precise description is often possible. We will first describe a quite general approach, due to Treschev, for which one only assumes the existence of a special normal form in the neighborhood of the torus, and which will allow us to give a canonical definition of the splitting matrix. Then, in the perturbative setting described in §1.9, we examine various usual definitions in connection with ours.

1.11.2. Normal forms and canonical splitting matrices. We begin with various results about normal forms in a neighborhood of an isotropic hyperbolic torus. The first one is new and apply to C^r systems, with $r \geq 3$, and is furthermore free from any arithmetical assumption on the rotation vector of the torus. We then describe the normal forms derived in [Gr] and [T1] in the analytic case and under the usual Diophantine assumption on the rotation vector. Finally we sketch the refinements obtained in [El] and [Ni3] for 1-hyperbolic tori. We simply hope that this short excursion will shed some geometric light on the situations and questions considered in Chapter 2.

The setting is as above with (M, Ω) a smooth 2ℓ-dimensional symplectic manifold, H a Hamiltonian on M, X_H its vector field and Φ the Hamiltonian flow. We assume that Φ leaves invariant an isotropic torus $\mathcal{T}$ contained in a *normally* hyperbolic invariant manifold. We furthermore assume that there exists a C^r symplectic coordinate system (ϕ, I, s, u) in which the torus is given by $\mathcal{T} = \{y = 0, u = s = 0\}$, the normally hyperbolic manifold by $\mathbf{N} = \{u = s = 0\}$, the invariant manifolds attached to $\mathcal{T}$ by $\mathcal{W}^+ = \{I = 0, u = 0\}$, $\mathcal{W}^- = \{I = 0, s = 0\}$ and finally the coisotropic invariant manifolds attached to $\mathbf{N}$ by $\mathbf{W}^+ = \{s = 0\}$ and $\mathbf{W}^- = \{u = 0\}$.

In such a coordinate system, one checks immediately that the Hamiltonian takes the following form:

(N) $\qquad H(\phi, I, s, u) = C(\phi) \cdot I + I \cdot D(\phi) I + u \cdot \Lambda(\phi) s + g(\phi, I, s, u),$

where $C \in C^r(\mathbb{T}^n, \mathbb{R}^n)$, $D \in C^r(\mathbb{T}^n, M_n(\mathbb{R}))$, $\Lambda \in C^r(\mathbb{T}^n, M_m(\mathbb{R}))$, and $g = O_3(I, s, u; \phi)$. Moreover one has the additional vanishing properties: $\partial_\phi g = 0$ on $\mathcal{T}$, $\partial_s g = 0$ on $\mathcal{W}^+$, $\partial_u g = 0$ on $\mathcal{W}^-$.

One can notice the similarity between the normal form (N) and the one given by S.Graff and D.Treshchev in the analytic case (see below). It can actually be proved that the main dynamical properties around a torus (λ-lemma, obstruction property in the nonresonant case, etc.) are the same for (N) or for the stronger form (GT) given below.

The underlying manifold M and the Hamiltonian H are now supposed to be analytic. An n-dimensional invariant torus $\mathcal{T}$ in M will be called a *Graff-Treschev torus* if there exists an analytic symplectic coordinate system (ϕ, I, s, u) in a neighborhood of $\mathcal{T}$ in $T^*(\mathbb{T}^m \times \mathbb{R}^n)$ in which the Hamiltonian function of the system takes the following form:

$$\text{(GT)} \qquad H(\phi, I, s, u) = \omega \cdot I + \frac{1}{2} I \cdot AI + \Lambda(\phi)\, u \cdot s + g(\phi, I, s, u),$$

where ω is a Diophantine vector in $\mathbb{R}^n$, A is a symmetric matrix of size n and $\Lambda(\phi)$ a ϕ-dependent matrix of size m whose eigenvalues have positive real parts, and finally where $g(\phi, I, s, u) = O_3(I, s, u; \phi)$. In such a coordinate system, the invariant torus is given by $\mathcal{T} = \{I = 0, u = s = 0\}$. Using classical hyperbolic KAM techniques as elaborated in [Gr], D.Treschev proved in [T1] that the tori appearing in near-integrable systems when a resonant Lagrangian torus is destroyed are indeed Graff-Treschev tori.

In the special but important case of a simple resonance, *i.e.* when $m = 1$, and still assuming analyticity of the data, it is possible to improve on the preceding normal form (GT), as was shown by H.Eliasson [El] and L.Niederman [Ni3]. Namely, given a Graff-Treschev torus with positive definite matrix A, there exists a neighborhood O of $\mathcal{T}$ and a symplectic coordinate system (ϕ, I, s, u) in O such that the system takes the following form:

$$\text{(EN)} \qquad H(\phi, I, s, u) = \omega \cdot I + \frac{1}{2} I \cdot AI + \lambda\, us + g(\phi, I, s, u),$$

where $\lambda > 0$ is a scalar and now the remainder g has the special form $g(\phi, I, s, u) = O_2(I, us; \phi, s, u)$. The invariant torus is given as before by $\mathcal{T} = \{I = 0, u = s = 0\}$, but now in addition its invariant Lagrangian manifolds are straightened: $\mathcal{W}^+ = \{I = 0, s = 0\}$, $\mathcal{W}^- = \{I = 0, u = 0\}$, and the flow on them is *linear* (with quasifrequency ω).

An important remark on these normal forms, due to Treschev, is the following rigidity result:

THEOREM.

1. *Let $\mathcal{T}$ be a Graff-Treschev torus in M, and let (ϕ, I, s, u), (ϕ', I', u', s') two symplectic coordinate systems in which the Hamiltonian takes the form* (GT). *Then:*
i) There exist $b \in \mathbb{T}^n$, $B \in GL_n(\mathbb{Z})$ (i.e. $\det B = \pm 1$) and a function $P: \mathbb{T}^n \to GL_d(\mathbb{R})$ such that:

$$\phi' = b + B\,\phi + O_2, \quad I' = {}^t(B^{-1})\,I + O_2, \quad u' = P(\phi)\,u + O_2, \quad s' = {}^t(P^{-1}(\phi))\,s + O_2,$$

where O_2 stands for $O_2(I, s, u; \phi)$;
ii) Denoting by Λ and Λ' the matrices appearing in (GT) *for the two coordinates systems respectively, one has:*

$$(\omega \cdot \frac{\partial}{\partial \phi})P + P\Lambda = \Lambda' P;$$

iii) If the eigenvalues of Λ are real and simple, the matrix P is constant and $\Lambda' = P\Lambda P^{-1}$.

2. Let $\mathcal{T}$ be a Graff-Treschev torus in M of dimension $n = \ell - 1$. Let (ϕ, I, s, u), (ϕ', I', u', s') be two symplectic coordinate systems in which the system takes the form (EN). Then there exist $b \in \mathbb{T}^n$, $B \in GL_n(\mathbb{Z})$ and $P \in \mathbb{R}^*$ such that:

$$\phi' = b + B\phi + O_2, \quad I' = {}^t(B^{-1})I + O_2, \quad u' = Pu + O_2, \quad s' = \frac{1}{P}s + O_2,$$

where O_2 stands for $O_2(I, s, u; \phi)$.

The first part of this theorem is proved in [T3], and the second one is just an easy consequence.

If $\mathcal{T}$ is a (EN) torus, Treschev gives in [T3] a nice construction of a symplectic invariant describing the splitting of the invariant manifolds $\mathcal{W}^\pm = W^\pm(\mathcal{T})$. The first remark is that there exists two naturally defined Lie subalgebras $\mathcal{L}^\pm$ of the algebras $\mathcal{X}(\mathcal{W}^\pm)$ of vector fields on $\mathcal{W}^\pm$: given a vector $a \in \mathbb{R}^n$, one proves the existence of uniquely defined vector fields $w^\pm(a)$ tangent to $\mathcal{W}^\pm$, satisfying

$$w^-(a) = a.\partial_\phi + O(u^2), \qquad w^+(a) = a.\partial_\phi + O(s^2);$$

where as usual $a.\partial_\phi = \sum_{k=1}^n a_k \partial/\partial\phi_k$. The maps $w^\pm : \mathbb{R}^n \to \mathcal{X}(\mathcal{W}^\pm)$ are injective Lie algebra morphisms ($\mathbb{R}^n$ being endowed with its trivial Lie algebra structure), and one sets $\mathcal{L}^\pm = w^\pm(\mathbb{R}^n)$. One can moreover notice that $\mathcal{L}^\pm$ are commutative; and that each vector field in $\mathcal{W}^\pm$ commutes with the Hamiltonian vector field X_H. Finally, given a point x in $\mathcal{W}^\pm$, one proves that $X_H(x)$ is *transverse* in $T_x\mathcal{W}^\pm$ to the n-dimensional vector space $V^\pm(x)$ generated by the vectors $X(x)$, $X \in \mathcal{L}^\pm$. All these results are proved in [T3], using only the weaker normal form (GT) (still in the $m = 1$ case), but they are indeed easy consequences of the existence of the normal form (EN), since the flow on the local invariant manifolds is linear.

Assume now that Γ is a homoclinic orbit biasymptotic to the torus $\mathcal{T}$, and fix a point $x \in \Gamma$. Then the two spaces $V^+(x)$ and $V^-(x)$ detect the nonvanishing part of the splitting of $\mathcal{W}^+$ and $\mathcal{W}^-$ at x. They possess canonical bases $\mathcal{B}^\pm = (w_1^\pm, \cdots, w_n^\pm)$, associated with the canonical basis of $\mathbb{R}^n$, and one can form the following quantity:

$$I(x) = \omega^n(w_1^+, \cdots, w_n^+, w_1^-, \cdots, w_n^-),$$

which is just the symplectic volume of the union of the two bases. The interest of the construction is that the quantity $I(x)$ has an intrinsic character : it does not depend on the choice of the point x on Γ, neither on the particular choice of a normal form coordinate system (due to the rigidity). In particular, the vanishing of I is equivalent to the nontransversality (in the energy level) of the intersection of $\mathcal{W}^+$ and $\mathcal{W}^-$ along Γ.

As a final but important remark, notice that Treschev's construction makes it possible to define *canonically* the splitting matrix at any point x of Γ, as soon as a normal form coordinate system for the torus is fixed. Indeed,

$$\mathcal{C}^- = (X_H, w_1^-, \cdots, w_n^-)$$

is a canonically defined basis for $T_x\mathcal{W}^-$, so we only need to fix a Lagrangian subspace $F \subset T_xM$ transverse to $T_x\mathcal{W}^-$ (*i.e.* the direction of the splitting that we want to consider) and the splitting matrix in the direction F is just the matrix of the splitting form of the pair $(T_x\mathcal{W}^-, T_x\mathcal{W}^+)$, relative to the decomposition $(T_x\mathcal{W}^-, F)$, expressed in the canonical basis $\mathcal{C}^-$.

1.11.3. Poincaré-Melnikov integrals and the splitting in the perturbative setting.
We now come to the splitting in the particular case of our perturbative Hamiltonian

$$H(\phi, I, q, p) = F(I, q, p) + \mu G(\phi, I, q, p),$$

with $(\phi, I) \in T^*\mathbb{T}^n$ and $(q, p) \in T^*\mathbb{T}^1$. Our aim here is to make precise the connections between our definitions and the usual notions introduced for measuring the splitting. As a reference we will use the definitions of [T2] and [DG2], based on Poincaré-Melnikov integrals, which are the most commonly used.

We keep the notation of §1.10.1, in particular we set again

$$\mathcal{T}_0 = \{(\phi, 0, 0, 0), \ \phi \in \mathbb{T}^n\}, \qquad \mathcal{W}_0^+ = \mathcal{W}_0^- = \{(\phi, 0, q, X(q)), \ \phi \in \mathbb{T}^n, \ q \in \mathbb{T}^1\}$$

for the unperturbed objects, and

$$\mathcal{T} = \left\{ (\phi, I(\phi), Q(\phi), P(\phi)), \ \phi \in \mathbb{T}^n \right\},$$
$$\mathcal{W}^\pm = \left\{ (\phi, I^\pm(\phi, q), q, P^\pm(\phi, q)), \ \phi \in \mathbb{T}^n, \ q \in D^\pm \right\}$$

for the perturbed ones. We will fix the parameter q_0 and consider the intersection of the invariant manifolds with the section $\{q = q_0\}$. Given $\phi \in \mathbb{T}^n$, we denote by γ_ϕ the solution of the unperturbed problem for the initial condition $(\phi, 0, q_0, X(q_0))$, i.e. for the point of angle ϕ in the intersection of the invariant manifolds with the section $\{q = q_0\}$; this is of course a homoclinic solution for the torus $\mathcal{T}_0$. Recall that the Poincaré-Melnikov integrals reads

$$L(\phi) = \int_{-\infty}^{+\infty} (G - \overline{G} - \{\chi, F\}) \circ \gamma_\phi(t) \, dt$$

where $\overline{G}$ stands for the average of the function $G(., 0, 0, 0)$ and where χ is a solution of the classical homological equation

$$(\omega . \partial_\phi) \chi = G(., 0, 0, 0) - \overline{G},$$

with $\omega = \partial_I F(0, 0, 0)$ (see [DG2]). As noticed by Treschev in a slighty different form, the interest of the term $\{\chi, F\}$ is to make the integral absolutely convergent. It is easy to prove that

$$\Delta I(\phi) = I^+(\phi) - I^-(\phi) = \mu \, \partial_\phi L(\phi) + O(\mu^2);$$

in other words $\partial_\phi L(\phi)$ measures the first order of the difference between the I-coordinates of the invariant manifolds in the section $\{q = q_0\}$. Assume now that ϕ_0 corresponds to a homoclinic point x, i.e. $\Delta I(\phi_0) = 0$. Then one defines the splitting matrix $S = [S_{i,j}]$ as the matrix of the splitting form of the pair $(T_x \mathcal{W}^-, T_x \mathcal{W}^+)$, relative to the canonical horizontal and vertical decomposition of $T_x(T^*\mathbb{T}^\ell)$, and expressed in the basis

$$e_0 = X_H, e_1 = \partial_{\phi_1}, \ldots, e_n = \partial_{\phi_n}.$$

The coefficients are easily derived:

$$S_{ij} = \mu \, \partial^2_{\phi_i \phi_j} L(\phi_0) + O(\mu^2), \qquad 1 \leq i, j \leq n;$$

while of course $S_{0,j} = S_{i,0} = 0$. One can also use Treschev's formalism (see [T3]) and write, for $1 \leq i, j \leq n$:

$$S_{ij} = \lim_{T \to +\infty} \Big(\int_{-T}^{T} \big\{ I_i, \{I_j, G\} \big\} \circ \gamma_\phi(t) \, dt \\ + \big\{ I_i, \{I_j, \chi\} \big\} \circ \gamma_\phi(T) - \big\{ I_i, \{I_j, \chi\} \big\} \circ \gamma_\phi(-T) \Big).$$

Needless to say, the definitions of the splitting that we will use in the sequel are those introduced here. In Chapter 2 we will consider the perturbative setting introduced in §1.9 and prove the *anisotropic* character of the splitting matrix for m-resonant tori. Chapter 3 is devoted to the analytical study of the splitting for Hamiltonian (∗), and in this case the invariant manifolds are exact, so the splitting matrix is directly expressed in terms of the derivative of the generating functions, as shown in §1.7.

CHAPTER 2

Estimating the Splitting Matrix Using Normal Forms

In Chapter 1 we developed the geometric side of the theory in a general setting, in order to have it available under a variety of circumstances, not necessarily of perturbative nature. Of course the theory can be successfully and indeed rather easily applied in the framework of *regular* perturbation theory where one can derive finite-order approximations and study all sorts of interesting phenomena. But in this chapter we go directly to the more difficult question of studying exponentially small splitting phenomena. As mentioned in the introduction the truly relevant case for what is usually called "Arnold diffusion", *i.e.* global instability of near-integrable Hamiltonian systems, is the case of analytic perturbations of convex (more generally steep) systems. Here we will address the problem of the evaluation of the splitting distance and splitting matrix using what can be called the *symplectic* method. In fact we essentially confine ourselves to the study of the homoclinic problem (splitting matrix) which in some ways is more difficult from the analytical viewpoint, whereas the heteroclinic problem requires some additional geometry, for which of course the results of Chapter 1 would be quite helpful.

Let us briefly review the methods and results. The first step consists in performing a large number of resonant normalizations, that is in using resonant normal forms to an exponentially high order, a technique introduced by N.N.Nekhoroshev, which we recall and use in §2.1. Having normalized the system we are interested in, we then perform in §2.2 a—more or less classical—computation in the vinicity of a resonant surface, which prepares the ground for finding hyperbolic invariant tori, homoclinic trajectories and so on. It also explains how the Hamiltonian (∗) naturally arises. In all this it is crucial to be able to keep track of the various quantities under canonical transformations since we ultimately want to estimate the splitting in a convenient coordinate system. For this we use the transformation formulas derived in Chapter 1 but the setting is now perturbative, which leads to simplifications. Then the fact that the splitting matrix is symmetric entails important stability properties, which are typical of self-adjoint operators. We pause in order to review these properties in §2.3 which explains how the geometric objects of Chapter 1 are reflected in the perturbative situation. This gives the setting which is used in Chapter 3 as well. Putting things together, we get in §2.4 a general but provisional statement on upper exponential bounds for the splitting matrix. Here "provisional" refers to the fact that we *assume* the existence of the invariant object (a torus) we are studying, together with its invariant manifolds and a family of homoclinic trajectories. We are not yet concerned with proving its existence, using *e.g.* KAM-type techniques. The bounds we derive are quite general, but also quite detailed in the sense that they are *local* and *anisotropic*. Let us phrase this informally, because it is perhaps the main "message" of this chapter; what we show can be loosely expressed by saying that *the splitting transverse to a resonant surface is generically polynomial whereas it is exponentially small along the surface*. This ties up nicely with the picture derived from the local stability exponents, as we make more precise in §2.6, where we also comment on the relationship which has emerged between splitting and stability in near-integrable Hamiltonian systems.

With hindsight the results of this paragraph are not so surprising but they do use the spectral stability of self-adjoint operators in order to control the eigenvalues and indeed also the eigenprojectors of the splitting matrix. In §2.5 we review cases in which the results of §2.4 can be effectively applied, recalling some classical KAM results and using (particular cases of) the results of Chapter 1.

This chapter is the first place in this paper where exponentially small quantities appear and it thus seems appropriate to fix some terminology. There is indeed a whole range of problems where such asymptotic problems occur, from one frequency linear to multifrequency nonlinear questions and we are dealing mostly with the latter where the situation is quite intricate, in the sense that many quantities in fact do *not* have an easily describable asymptotic behaviour (see below).

So let us introduce some convenient definitions: let $\eta(\varepsilon)$ be a positive scalar quantity, defined and continuous for $\varepsilon \geq 0$ small enough, with $\eta(0) = 0$. We say that η is *exponentially small* if $\exp(\varepsilon^{-\alpha})\eta(\varepsilon)$ is bounded (for small nonzero ε) for some $\alpha > 0$. If so define a ($0 \leq a \leq \infty$) to be the supremum of such α's. In other words if one sets:

$$E(\eta) = \{\alpha \geq 0, \ \eta(\varepsilon) = o\big(\exp(-\big(\frac{1}{\varepsilon}\big)^\alpha)\big)\},$$

then a is the least upper bound of $E(\eta)$.

Now if $a < \infty$ we say that $\eta(\varepsilon)$ is exponentially small with *exponent* $a = a(\eta)$. In this case we can further define the *width* $w = w(\eta) \geq 0$ much in the same way: it is the supremum of those $c > 0$ such that $\eta(\varepsilon) = O\big(\exp(-c\big(\frac{1}{\varepsilon}\big)^a)\big)\}$. We propose to call the constant defined in this way the "width" because it is often connected (rigorously or conjecturally) with the width of the analyticity strip of some function. Here we have $0 \leq w \leq \infty$ with w constant, but in some situations one needs in fact to consider a nonconstant width $w(\varepsilon)$ defined as some elementary asymptotic equivalent of $\varepsilon^a \log\big(\frac{1}{\eta(\varepsilon)}\big)$. These things happen in nature in fact, due for instance to the fact that entire functions have infinite analyticity width (see [S], [DGJS] and §3.5 below).

If w (constant or not) is finite, one may inquire about the behaviour for small ε of the quantity $\exp(w\varepsilon^{-a})\eta(\varepsilon)$ which is called the *prefactor*. This may or may not be "well-behaved" as a function of ε and in fact we will be interested in a case (the Melnikov integral in the multifrequency case) where it is surely very complicated to analyze (see especially §3.5 below). Lastly one may substract the thus determined leading term and investigate the screened off exponential factors (with smaller exponents), of whatever the remainder might look like; one can also repeat these operations infinitely many times, giving rise to transseries etc. But we are very far from understanding such fine behaviours in the kind of problems we are dealing with.

In fact this terminology being set, it should be stressed that in this chapter we derive only *upper* bounds for the splitting (essentially the splitting matrix). These are probably sharp as far as exponents are concerned, but we do not keep track of the widths, so that we certainly do not get lower bounds. This is a fundamental feature of the symplectic method as is to-date; namely in the first step (normalization) we already basically lose track of the width and it is a major puzzle to understand how this could be repaired. The symplectic method presented here is more flexible and general than the *analytic* method of Chapter 3, it reflects the geometry of resonances quite well but it is not as precise. We will see more of this at the end of Chapter 3, after having developed both approaches.

NOTE. In what follows we make use of well-established tools of canonical perturbation and try to emphasize the simple concepts at work. We do not insist on estimating the various constants that occur and use compact notation. In particular, if A and B are two scalar quantities depending on a parameter, $A \preceq B$ means that there exists a constant c such that $A \leq cB$, i.e. $A = O(B)$. The meaning should be obvious from the context, including the set of parameters on which the implied constant c depends. We use $\succeq$ and $\asymp$ in a similar fashion. If we still need to give names to some constants we normally use generic letters, so that two constants with the same name may be different. We insist nevertheless that almost everything we do is potentially explicit, i.e. the constants could be estimated if need be.

2.1. Resonant normal forms

So to start with, we consider the familiar near-integrable system:

$$(1) \qquad H(I,\phi) = h(I) + \varepsilon f(I,\phi), \qquad (I,\phi) \in P \times \mathbb{T}^\ell \quad (P \subset \mathbb{R}^\ell).$$

Here we normalize the circle to have length 2π, that is we set $\mathbb{T} = \mathbb{R}/2\pi\mathbb{Z}$, so that f is given by a Fourier series:

$$(2) \qquad f(I,\phi) = \sum_{k \in \mathbb{Z}^\ell \setminus \{0\}} f_k(I) e^{ik \cdot \phi}.$$

Note that we have omitted the constant corresponding to $k = 0$, which means that we lump the average of the perturbation with the unperturbed part but often omit for typographical simplicity the mention of the dependency on ε of the various functions (here h and f). We set $\omega(I) = \nabla h(I)$ for the frequency map and assume ω does not vanish on the nice (i.e. convex compact) I-domain P we are considering; this amounts to the fact that the unperturbed system has no fixed point. Note that below the notation does not exactly follow the one in the introduction for Hamiltonian $(*)$, because the setting is more general and in particular we do not single out a hyperbolic degree of freedom corresponding to the pendulum variables (p,q). These minor notational discrepancies should hopefully cause no confusion.

We assume that h and f are real *analytic* and since P is compact the function H actually admits a continuation to a complex strip of finite width. For a domain $D \subset \mathbb{R}^\ell$ and any $r > 0$, we let D_r denote the union of the balls of radius r centered at points of D. For simplicity we do not always specify the complex extensions, especially as they do not appear in the statements of the results. Analyticity is required here because we are interested in *exponentially* small estimates for the splittings. However much of what is done below can be adapted to other regularity classes, in particular finitely differentiable functions or C^∞ functions with some Gevrey growth conditions. Some of these variants deserve interest, in particular the interaction of Gevrey conditions with arithmetic properties in normal forms.

2.1.1. We first recall a few useful things about resonant normal forms, keeping this reminder to a strict minimum. Actually for our immediate purpose, namely for the study of *homoclinic* splitting problems, we will need a local lemma only, and for ease of reference we will use the statement in [Pöl] (p. 192; see also the remarks there). We have gathered some remarks in §2.1.5 below.

Let $\mathcal{M} \subset \mathbb{Z}^\ell$ be a submodule of rank m with $0 \leq m < \ell$. If $m > 0$, $\mathcal{M}$ corresponds to a resonance of multiplicity m; we take $m < \ell$ because we have assumed that ω does not vanish on P. In this situation we set $n = \ell - m$ and we let $\Pi_\mathcal{M}$ denote the n-plane orthogonal to $\mathcal{M}$ in the ℓ-dimensional frequency ω-space; then we let $S_\mathcal{M} = \omega^{-1}(\Pi_\mathcal{M}) \subset P$ be the pull-back of $\Pi_\mathcal{M}$ in action I-space. Explicitly:
$$S_\mathcal{M} = \{\, I \in P \mid \omega(I) \cdot k = 0 \quad \text{for all } k \in \mathcal{M}\,\}.$$
Here we do *not* make any assumption on the frequency map $I \mapsto \omega(I)$. In general the set $S_\mathcal{M}$ can be anything, including possibly the whole of the domain P in action space; think of the case when the frequency map is constant, corresponding to a linear h describing a set of non-interacting harmonic oscillators. Now of course when the frequency map is a local diffeomorphism, *i.e.* when h is nondegenerate (or fully nonlinear or anisochronous depending on the terminology), $S_\mathcal{M}$ classically defines the resonant surface associated to $\mathcal{M}$ and is indeed a smooth n-dimensional surface in action space. But the results in this paragraph do not depend on h being nondegenerate.

Given such an integer lattice $\mathcal{M}$, one defines the corresponding (resonant) normal forms as follows:

DEFINITION. *Let $\mathcal{M} \subset \mathbb{Z}^\ell$ be a submodule of rank m with $0 \leq m < \ell$. A Hamiltonian H as in Equation (1) is said to be in normal form with respect to $\mathcal{M}$ (or in $\mathcal{M}$-resonant normal form) if in the expansion (2) one has $f_k = 0$ for $k \notin \mathcal{M}$.*

Note that this is the same as requiring that $\partial f / \partial \phi \in \mathcal{M} \otimes \mathbb{R}$ where $\mathcal{M} \otimes \mathbb{R}$ is the m-plane generated by $\mathcal{M}$ in *action* I-space. Thinking of the equations of the motion for the action variables (namely $dI/dt = -\varepsilon \partial f / \partial \phi$), this formulation gives the key to the dynamical meaning of the normal forms: $\omega \cdot I$ will then be a first integral if $\omega \in \Pi_\mathcal{M}$.

2.1.2. Let now $\alpha > 0$ be a—small—constant and $K > 0$ be a—large— constant; both may and eventually will depend on the perturbation parameter ε. Following [N] we have the

DEFINITION. *A domain $B \subset \mathbb{R}^\ell$ in ω-space is (α, K)-nonresonant modulo $\mathcal{M}$ if for all $\omega \in B$,*
$$|\omega \cdot k| \geq \alpha \quad \text{for all } k \in \mathbb{Z}^\ell \setminus \mathcal{M} \text{ such that } |k| \leq K.$$
A domain D in action I-space is called (α, K)-nonresonant modulo $\mathcal{M}$ if $\omega(D)$ is.

If D is (α, K)-nonresonant modulo $\mathcal{M}$, one can transform the Hamiltonian H to a $\mathcal{M}$-resonant normal form on some neighbourhood D_r of D up to a small remainder. This is the content of the Normal Form Lemma ([Pö1], p. 192), which is valid for any values of the parameters α, K and does not depend on any nondegeneracy assumption on the integrable part h. As it is really the crux of the matter here, we first state it without the constants (which appear explicitly in [Pö1]) as:

PROPOSITION. *Suppose D is contained in the interior of P and is (α, K)-nonresonant modulo $\mathcal{M}$ for some lattice $\mathcal{M}$. Then for ε small enough, there exists a canonical transformation $\mathcal{C}_\varepsilon$ which is ε-close to identity, is defined over D_r with $r \asymp \alpha/K$ and such that over D_r the transformed Hamiltonian $H \circ \mathcal{C}_\varepsilon$ is in resonant normal form up to a remainder of size $O(\exp(-cK))$ for some constant $c > 0$.*

Since D is assumed to be contained in the interior of P, one has indeed that $D_r \subset P$ for r small enough. Note that with respect to [Pö1], we changed r into $2r$ for ease of notation; since we are not dealing with the precise values of the involved constants, this is of course immaterial.

2.1.3. We will use this statement for a fixed given $\mathcal{M}$ only and so make it more explicit then. Fixing $\mathcal{M}$ we can perform a linear symplectic change of coordinates which transforms this lattice into standard form: namely we can assume that $\mathcal{M}$ is generated by the m last basis vectors in $\mathbb{R}^\ell$ (*cf.* §2.1.5 for some remarks on this algebraic operation). Now write $I = (I_1, I_2)$, $\omega = (\omega_1, \omega_2)$ etc. with vectors of sizes n and m, $\ell = n + m$ (again with a slight departure from the introduction; the value of m is shifted by 1). The conclusion in Proposition 2.1.2 can then be rephrased and detailed as:

PROPOSITION. *Under the assumptions of §2.1.2 there exists for small enough ε a canonical transformation $\mathcal{C}_\varepsilon$ which is ε-close to identity, defined over D_r and such that over this domain:*

$$(3) \quad H \circ \mathcal{C}_\varepsilon(I, \phi) = h(I) + \varepsilon Z(I, \phi_2, \varepsilon) + R_\varepsilon(I, \phi, \varepsilon) = H_r(I, \phi_2, \varepsilon) + R_\varepsilon(I, \phi, \varepsilon).$$

Here Z is bounded, so that the resonant part H_r is ε-close to the original integrable part h, and the remainder R_ε is estimated as: $R_\varepsilon \preceq \exp(-cK)$.

Here we have explicitly emphasized the dependence of the resonant part Z on ε and we record for future use the fact that the dominant term can be immediately written down:

$$(4) \quad Z(I, \phi_2, 0) = \sum_{k_2 \in \mathbb{Z}^m \setminus \{0\}} f_{0,k_2}(I) e^{ik_2 \cdot \phi_2},$$

with the decomposition $k = (k_1, k_2)$ as usual.

We will eventually pick $K \asymp \varepsilon^{-a}$ with some exponent $a > 0$, so that R_ε will indeed be exponentially small with respect to ε. In these coordinates the resonant set $S_\mathcal{M}$ is defined by $\omega_2 = \omega_2(I) = \partial h / \partial I_2 = 0$. If h is nondegenerate, the n first variables I_1 give a local parametrization of the surface $S_\mathcal{M}$, whereas the m last variables I_2 are transverse to this resonant surface. The normal form (3) expresses that under the assumptions of Proposition 2.1.2 the nonresonant angles ϕ_1 can be transformed away up to an exponentially small remainder R. Forgetting about R, the variables I_1 provide n prime integrals for the resonant part H_r, which can be viewed as a suspension of the m-dimensional system with fixed I_1 by the tori $I_1 = \mathrm{cst}$ spanned by the ϕ_1. Note that if $m = 1$, *i.e.* if we start from a simple resonance, H_r is completely integrable, a particularly simple but important case to be further discussed below.

The normal form (3) carries important information about the splitting and the rest of this chapter is basically spent unraveling this information which was expressed in the italicized motto of its introduction. The point is that in the system governed by H_r the splitting matrix for *any* invariant object, or for that matter the distance between the perturbed and unperturbed manifolds, will vanish in the I_1 direction, *i.e. along* the resonant surface. Reinstating the remainder will produce an exponentially small splitting in these directions, which for the splitting matrix materializes as n exponentially small eigenvalues.

2.1.4. We now have to find situations where (α, K)-nonresonant domains present themselves, with specific values for α and K. Although more general applications of the same techniques are certainly conceivable, we will apply the above proposition in the vicinity of invariant tori for h. We start with a Diophantine n-vector ω_1, i.e. such that:

(5) $\quad |\omega_1 \cdot k_1| \geq \gamma |k_1|^{1-\tau_1} \quad$ for all $k_1 \in \mathbb{Z}^n \setminus \{0\}$, with some constants $\gamma > 0, \tau_1 \geq n$.

By definition, i.e. by (5), for any $K \geq 1$ the frequency vector $\omega = (\omega_1, 0)$ is $(\gamma K^{1-\tau_1}, K)$-nonresonant modulo $\mathcal{M}$ (which is spanned by the last m basis vectors as above). This requires a one-line justification. Namely if $k = (k_1, k_2)$ is not in $\mathcal{M}$ and satisfies $|k| < K$, then k_1 is not 0 and $|k_1| \leq K$; then apply (5).

More generally it is useful to be able to let the frequency of the unperturbed torus vary. We thus extend the above as follows: let $\omega(\varepsilon)$ be a smooth family of frequencies defined for ε small enough, with $\omega(0) = \omega = (\omega_1, 0)$ as above, and such that $\omega(\varepsilon) = \omega(0) + O(\sqrt{\varepsilon})$. A case of particular interest, corresponding to 1-hyperbolic tori lying in the vicinity of an m-fold resonance, is given by: $\omega(\varepsilon) = (\omega_1, \sqrt{\varepsilon}\omega_2, 0)$ with ω_2 a Diophantine $(m-1)$-vector (this $m-1$ would become m in the notation of the introduction, corresponding to the model $(*)$). Then we would like that $\omega(\varepsilon)$ be, say, $(\alpha/2, K)$-nonresonant modulo $\mathcal{M}$, with (α, K) as above for ω, i.e. $\alpha = \gamma K^{1-\tau_1}$. In other words we lose at most a factor 2, say, by shifting the frequency somewhat. Again let us compute, with $k \in \mathbb{Z}^\ell \setminus \mathcal{M}$:

(6) $\quad\quad\quad |\omega(\varepsilon) \cdot k| \geq |\omega(0) \cdot k| - c\sqrt{\varepsilon} K \geq \alpha - c\sqrt{\varepsilon}K,$

for some constant c. So that, dropping constants, we need to have: $\alpha \succeq \sqrt{\varepsilon}K$. Although we will not provide all the (rather straightforward) details, this says in essence that the symplectic technique we use in this chapter is flexible enough to study the approach of a resonance of given multiplicity m along resonances of any lower multiplicities, here along simple resonances: the family $\omega(\varepsilon) = (\omega_1, \sqrt{\varepsilon}\omega_2, 0)$ indeed describes the vinicity of an m-fold resonance on a simply resonant surface (see also §3.6.3 below).

Coming back to the case of a constant frequency vector, these simple but essential computations suggest the following choice of parameters:

(7) $\quad\quad\quad \alpha \asymp K^{1-\tau_1}, \quad r \asymp \alpha/K, \quad K \asymp \varepsilon^{-a}, \quad$ with $a^{-1} = 2\tau_1.$

We encapsulate these conclusions in the

PROPOSITION. *Let ω_1 be a Diophantine n-vector satisfying inequalities (5). Assume that the origin of the I coordinates is chosen so that $\omega|_{I=0} = (\omega_1, 0)$ and pick the parameters as in (7). Then Proposition 2.1.3 applies with D reduced to the origin $\{I = 0\}$, and so with D_r the sphere of radius r centered there.*

One has $r \asymp \sqrt{\varepsilon}$ and for any $\rho > 0$ one can ensure $r \geq \rho\sqrt{\varepsilon}$ by picking the implied constant in K small enough, i.e. $K \leq c\varepsilon^{-a}$ with c small enough. So one gets in (3) a remainder R_ε which satisfies the estimate: $R_\varepsilon \preceq \exp(-w\varepsilon^{-a})$ for some $w > 0$.

In words, one can normalize Hamiltonian (1) over a sphere of radius r centered at 0 and of radius at least $\rho\sqrt{\varepsilon}$, where ρ is arbitrarily large and the remainder in (2) satisfies $R_\varepsilon \preceq \exp(-w\varepsilon^{-a})$ with $a^{-1} = 2\tau_1$. The second part of the proposition can be summarized by saying that $r/\sqrt{\varepsilon}$ can be made arbitrarily large at the expense of decreasing the width in the remainder, but keeping the exponent fixed.

2.1.5. We summarize here a few observations which should help put the above in perspective. We phrase them in an informal somewhat "physical" way, but they are really basic to the whole problem. What we have done above is to apply the Normal Form Lemma in the vicinity of a point which lies on a specified resonant set but is otherwise relatively Diophantine in the sense that it satisfies inequalities (5). We did that in such a way that the range of validity of the resonant normal form is at least $\rho\sqrt{\varepsilon}$ for an arbitrary large constant ρ. In fact this will enable us to work henceforth only over this domain. In other words, from now on we only need to consider Hamiltonian (3), with the data as in (5) and Proposition 2.1.4.

This can be made more transparent by a simple "dimensional analysis" which should also help understand the rest of this paragraph: the physical "resonance width" has order $\sqrt{\varepsilon}$; note that this is the natural perturbation parameter, rather than ε, because we are dealing with second order equations ($f = ma \ldots$). When starting from an n-dimensional torus with frequency $\omega = (\omega_1, 0)$ as above lying on an m-fold resonant surface, it will persist under suitable assumptions (including in particular nondegeneracy conditions) by KAM-theory, and will be shifted by an amount which is $O(\varepsilon)$. This being very small with respect to $\sqrt{\varepsilon}$, one may give some leeway and generalize to families with frequencies $\omega(\varepsilon)$ as above. Now if the tori which are born are hyperbolic, the corresponding Lyapunov exponents (transverse hyperbolic spectrum) will be $O(\sqrt{\varepsilon})$ and generically of that size. So the stable and unstable manifolds will depart from the torus from a quantity of that size in terms of action, over any finite "chunk" (*i.e.* in the compact-open topology; this can and will be made precise). We will investigate the splitting of homoclinic trajectories, and so these will also lie within $O(\sqrt{\varepsilon})$ of the unperturbed torus.

The upshot of all this is that we are indeed interested in a patch of size $O(\sqrt{\varepsilon})$, but it is crucial to be able to choose it as large as necessary on that scale, that is with the notation above to pick ρ arbitrarily large. In turn this can be done but at the expense of decreasing the width w in the remainder. This is absolutely crucial and the point where the symplectic method developed in this chapter is not as accurate (although more flexible and general) as the analytic method of Chapter 3. There is no way we can really control the width of the remainder, hence of the splitting, when effecting the series of normalizing transformations leading to the Propositions 2.1.2, 2.1.3 and 2.1.4 (see more on this in the closing paragraph §3.7.4).

Let us return to the role of the nature of the unperturbed Hamiltonian h. Again, what we have done up to now does not require any nondegeneracy condition. In fact the estimate on the radius r of validity of the Normal Form Lemma only *improves* if h is degenerate. This radius expresses that if a given point is (α, K)-nonresonant modulo $\mathcal{M}$ one can normalize over a ball in *frequency* space near that point. Pulling back to action space, the preimage of this ball contains a ball of proportional radius, simply because the frequency map is of class C^1. But it can actually contain much more, so that in fact one has $r \succeq \sqrt{\varepsilon}$ in general and $r \asymp \sqrt{\varepsilon}$ if h is nondegenerate. If on the contrary the frequency is a constant (h is linear), one can actually take a constant value r_0 for r (see *e.g.* Theorem 5 in [Pöl] and its short proof).

We will especially concentrate on the nondegenerate case, because we will need to apply some KAM-type results which do require nondegeneracy, *i.e.* the existence of torsion, although the results in the appendix would allow to weaken the assumptions in certain cases. But one can include degenerate cases, for instance a linear part h, provided one knows from some source or simply from the choice of the perturbation that the objects we are ultimately interested in (tori and homoclinic trajectories) exist. We will not discuss these relatively easy cases here but these isochronous or mixed cases are taken up in part in Chapter 3, albeit in particular cases. It is quite important to note that we will require nondegeneracy in order for KAM results to apply but *not* convexity or quasiconvexity or steepness, the assumptions under which the action variables are stable over exponentially long times. The evaluation of the splitting from above performed here is disconnected from these conditions, although in the convex case, say, there is a strong link between the speed of drift and diffusion and the size of the splitting, which is reviewed in [L4]. There is no contradiction of course here and *an exponentially small splitting certainly does not exclude a polynomially fast drift*. This does indeed happen in case the integrable part h is nondegenerate and not steep, for instance an indefinite quadratic form in the action variables; in terms of Hamiltonian (∗) the relative *signs* of the components of α does affect stability, but does *not* affect the size of the (homoclinic) splitting.

We now say a few words about the connection of the local reasoning above with the global geometry of resonances. We used only the Normal Form Lemma, which is classical and lies at the heart of the analytic part of Nekhoroshev's proof of the stability theorem. In that respect, [Pöl] is quite similar to [N]; the difference between the two papers lies in the geometric part, more precisely in the refined version of "Nekhoroshev's puzzle" which occurs in [Pöl]. We refer to the introduction of Section 4 in [Pöl] for details about this difference. The points we want to make, in connection with this covering of frequency space by blocks $B_\mathcal{M}$ associated with lattices $\mathcal{M}$ are the following (lattices are denoted by the letter Λ in [Pöl]). First this puzzle can be constructed in frequency space and then pulled back to action space. That is one covers frequency space by blocks $B_\mathcal{M}$ for varying $\mathcal{M}$ (we need not give the precise definitions here; see [N] and [Pöl]) and pulls back the pieces *via* the map $\omega = \nabla h : I \mapsto \omega(I)$, covering action space by the $\omega^{-1}(B_\mathcal{M})$'s. This construction does not require any nondegeneracy condition on h, but of course in the end does not guarantee long time stability unless h is steep. Assuming nondegeneracy (but not steepness), the puzzle in action space is a slightly distorted image of the one in frequency space because the frequency map is then a local diffeomorphism.

Here we have used just one given lattice $\mathcal{M}$ and have not considered all lattices simultaneously; it is thus possible to reduce our fixed lattice to a standard form *via* a linear symplectic change of coordinates, which is rather innocuous when considering just one lattice. But this reduction involves some lattice-theoretic constants which do play an important role when looking more globally and in particular affect the width of the exponentially small quantities. The computations involved are however purely algebraic and exact (in particular optimal). For details we refer to [L2] (Section III) and to [Pöl] (Section 4).

Another important observation in connection with the above is as follows: for $\tau_1 > n$ in (5), the point $\omega = (\omega_1, 0)$ does *not* lie, for ε small enough, in the block $B_\mathcal{M}$ associated to $\mathcal{M}$. We do not give the exact definitions because they are quite cumbersome, but the reader should hopefully be able to follow the idea below (or look up first into [N] and [Pöl]). The point is that as soon as $\tau_1 > n$ the point lies "too close" to other resonances; more precisely, for ε small enough, the point ω lies in $B_\mathcal{N}$ where $\mathcal{N}$ contains $\mathcal{M}$ strictly and is thus of rank strictly greater than $m = \text{rk}(\mathcal{M})$ (resonant modules can be taken to be primitive, that is they do not strictly contain any module of the same rank). What we have done is to apply a normal form lemma keeping in mind the requirement that it has to be valid over a domain of size $\sqrt{\varepsilon}$. This is *very important*, since ultimately the exponent a of the remainder, given by $a^{-1} = 2\tau_1$, also gives the exponent we derive for the homoclinic splitting. This will be further discussed in §2.6.6 in connection with stability properties.

Finally we note that the above does not generally apply to heteroclinic splitting problems, which are crucial in the study of instability but which we do not take up here. It would require a more refined discussion of the puzzle because *both* invariant objects which are being connected should be included in one and the same piece where a normal form is valid. This may be an interesting and many-faceted question.

2.2. Computations in the vicinity of a resonant surface

2.2.1. Having normalized the original system to the form (3) over the domain we are interested in, we first study the resonant part H_r from a rather general viewpoint. We start by effecting a convenient scaling. Set:

$$(8) \qquad I = \sqrt{\varepsilon} I', \quad \phi = \phi', \quad H = \varepsilon H', \quad t = t'/\sqrt{\varepsilon}.$$

This transformation multiplies the symplectic form by $\sqrt{\varepsilon}$ and preserves the equations of the motion. From (3) we get:

$$(9) \qquad H_r'(I', \phi') = \varepsilon^{-1} h(\sqrt{\varepsilon} I') + Z(\sqrt{\varepsilon} I', \phi_2', \varepsilon),$$

and this is defined over a ball $|I'| < \rho$, where ρ can be taken arbitrarily large by the second part of Proposition 2.1.4. We expand h as:

$$(10) \qquad h(I) = \omega_1 \cdot I_1 + \frac{1}{2} A I_1^2 + \frac{1}{2} B I_2^2 + C I_1 \cdot I_2 + O(I^3).$$

The notation is unambiguous for the A and B terms because A and B are square symmetric matrices of size n and m respectively. The mixed quadratic term is described by C which is an $m \times n$ matrix. We thus find that:

$$(11) \qquad H_r'(I', \phi') = \frac{\omega_1}{\sqrt{\varepsilon}} \cdot I_1' + \frac{1}{2} A I_1'^2 + \frac{1}{2} B I_2'^2 + C I_1' \cdot I_2' + Z(0, \phi_2', 0) + O(\sqrt{\varepsilon}).$$

Recall that H_r is independent of ϕ_1 so that I_1 is in fact a constant n-vector of the motion. So above I_1' is in fact constant. Note also that (4) gives an explicit formula for $Z(0, \phi_2, 0)$.

Assume now that the matrix $B = \partial^2 h / \partial^2 I_2$ is nondegenerate. Then we can shift the variable I_2' in (11) by the *constant* vector $B^{-1} C I_1'$, in other words we can set:

$$(12) \qquad (I_1'', I_2'') = (I_1', I_2' + B^{-1} C I_1').$$

Then (11) becomes:

$$(13) \quad H_r''(I'', \phi'') = \frac{\omega_1}{\sqrt{\varepsilon}} \cdot I_1'' + \frac{1}{2} A I_1''^2 - \frac{1}{2} C I_1'' \cdot B^{-1} C I_1'' + \frac{1}{2} B I_2''^2 + Z(0, \phi_2'', 0) + O(\sqrt{\varepsilon}).$$

The shift from I_2' to I_2'' allows to get rid of the mixed term in H_r but this does *not* immediately carry over to the full Hamiltonian H, because then I_1 is no more a constant of the motion, and (12) lifts to a symplectic transformation in the second set of variables, considering I'_1 as a parameter, but not to a symplectic transformation in the full set of variables.

Changing A to $A - {}^tCB^{-1}C$ and dropping primes and double primes, we have reduced the resonant partially integrable part H_r to the form:

$$(14) \quad H_r(I,\phi) = \frac{\omega_1}{\sqrt{\varepsilon}} \cdot I_1 + \frac{1}{2}AI_1^2 + \frac{1}{2}BI_2^2 + g(\phi_2) + O(\sqrt{\varepsilon}) = H_r^0(I,\phi_2) + O(\sqrt{\varepsilon}),$$

with the following assumptions and notation:

$$\begin{cases} (I_1, I_2) \text{ stands for } (I''_1, I''_2) \text{ (see (8) and (12))}; \\ B = \partial^2 h/\partial^2 I_2^2(0,0) \text{ is assumed to be invertible;} \\ g(\phi_2) = Z(0,\phi_2,0) \text{ is given by (4)}; \\ (14) \text{ is valid over a ball } \|I\| < \rho \text{ with } \rho \text{ arbitrarily large.} \end{cases}$$

We emphasize that in (14) the remainder is also independent of ϕ_1 (see (9)), so that I_1 is a constant of the motion. It is also important to note that $H_r^0(I,\phi_2)$ is independent of ε and is *not* the pertubation of an integrable system. Rather it can be seen as a generalized parametrized (by I_1) multidimensional "pendulum", which of course cannot be fully understood in general (for $m > 1$) because g is also arbitrary (see (4)).

The simple but enlightening computations above are not new in essence, although they are not easily to be found in this generality. We refer in particular to [N] (§11.6); see also [C] (§3.2 *sqq.*) for a more phenomenological viewpoint. We note that A.Neishtadt devoted a special study to the case with two degrees of freedom, including non-Hamiltonian systems; details and references are provided in [LM] (Chapter 4). The computation for a simple resonance ($m = 1$) appears in [El], [Ni3] and [DG1], to quote some recent papers. This "universal" computation also serves to provide a partial justification for the study of Hamiltonian (∗), which gives a model of the behaviour of a general near-integrable system (1) near a *simple* resonance. Note that (∗) does not contain the coupling term corresponding to the matrix C above. This term appears explicitly in [El] and [DG1] for the case of a simple resonance and also in [N] §11.6 in all generality, although in slight disguise.

2.2.2. As mentioned above, apart from the case of a simple resonance ($m = 1$), there is no hope to unravel $H_r^0(I,\phi_2)$ completely. But one can look for fixed points and these will give rise, for any given value of I_1, to n-dimensional m-hyperbolic tori $O(\sqrt{\varepsilon})$-close to $I_2 = 0$ for H_r, if ε is small enough. This can then be translated into the original variables, which amounts to a shift in I_2 and a scaling (see (12) and (8)).

More precisely, consider the "generalized pendulum" which derives from the Hamiltonian $P(I_2,\phi_2) = \frac{1}{2}BI_2^2 + g(\phi_2)$. We have already assumed that the matrix B is nondegenerate, and we now have to add the assumption that it is sign definite (as a quadratic form), say $B > 0$ for definiteness. The function $g : \mathbb{T}^m \to \mathbb{R}$ possesses a maximum ϕ_2^0 and we will need this maximum to be nondegenerate. This gives a nondegenerate hyperbolic fixed point x_0 for the pendulum P. Under these assumptions and for any given value I_1^0 ($|I_1^0| < \rho$), the n-torus spanned by the ϕ_1-variables, with equation $I_1 = I_1^0, I_2 = 0, \phi_2 = \phi_2^0$, is invariant and hyperbolic for the truncated Hamiltonian H_r^0 in (14). Reinserting the $O(\sqrt{\varepsilon})$ remainder, we see that this torus persists for ε small enough with a $O(\sqrt{\varepsilon})$ distorsion in the (I_2,ϕ_2) variables, because the remainder is still ϕ_1-independent, so that we are dealing with the perturbation of a nondegenerate hyperbolic fixed point for given I_1. We summarize all this in the

PROPOSITION. *Assume that the symmetric $m \times m$ matrix $B = \partial^2 h/\partial I_2^2(0,0)$ is nondegenerate and strictly positive as a quadratic form. Assume that the function $g(\phi_2) = Z(0, \phi_2, 0)$ defined by (4) has a nondegenerate maximum on $\mathbb{T}^m$. Then there exists a nondegenerate hyperbolic fixed point x_0 for the Hamiltonian P, which for any I_1^0 with $|I_1^0| < \rho$ and for ε small enough determines an n-dimensional m-hyperbolic torus $\mathcal{T}_\varepsilon(I_1^0)$ for Hamiltonian H_r (see (14)). It lies in the plane $I_1 = I_1^0$ and its projection on I_2-space is $O(\sqrt{\varepsilon})$-close to $I_2 = 0$.*

To be explicit, we find a smooth family of tori $\mathcal{T}_\varepsilon(I_1^0)$ with equations:

$$\begin{cases} I_1 = I_1^0, & \phi_1 \in \mathbb{T}^n, \\ I_2 = I_2(I_1^0, \sqrt{\varepsilon}), & \phi_2 = \phi_2(I_1^0, \sqrt{\varepsilon}). \end{cases}$$

This is immediately translated in terms of the original variables of Hamiltonian (1) by means of (8) and (12). In these original variables the distorsion is of size $O(\varepsilon)$ (see (8)), and the whole analysis is valid over a ball of radius $\rho\sqrt{\varepsilon}$ centered at the origin in I-space, with ρ arbitrarily large. This is because (12) does introduce a distorsion which depends on B and C, but since ρ can be taken arbitrarily large, it has no serious effect on the range of validity. We recall that all this leeway is provided by the fact that we are not trying to get a sharp estimate on the width of the exponentially small quantities which occur and deal only with the exponents in this section (see Proposition 2.1.4). The proposition above features the simplest possible statement in the situation at hand, but it is actually difficult to go beyond. More elaborate statements are conceivable at this point but they do not seem to be very useful yet because in the end we will have to reinsert the exponentially small non-integrable remainder R_ε in (3). We will elaborate on the geometric picture in §2.5, using in particular the results of Chapter 1. Note that we are working here in the setting of §§1.9–1.10, with a more analytic and quantitative flavor.

2.3. Splitting in a perturbative setting, variance and stability

Clearly if one keeps only the resonant part H_r in (3), the splitting of the invariant manifolds associated to any torus (or possibly a more general object) will simply vanish in the I_1 direction since these variables are constants of the motion for the flow generated by H_r. We now want to be able to perturb this situation, by taking the exponentially small remainder R_ε into account. Also, in order to put the Hamiltonian (1) in the normal form (3) we had to perform a near-identity canonical transformation $\mathcal{C}_\varepsilon$ and we want to be able to say something about the splitting in the *original* variables of (1). We address these questions in §§2.4–2.6; as for now we prepare the ground by specializing various notions introduced in Chapter 1 to the perturbative setting and see how they are related in that restricted situation. We also need some results from *linear* perturbation theory, which we gather in the first two sections.

2.3.1. Let us begin by reviewing some specific properties of self-adjoint operators, in a form tailored to our needs; we are dealing here with the simple finite-dimensional case, whereas the theory largely aims at infinite-dimensional situations. Due to the symmetry of the various "angular operators", these stability properties play an important role in the problem of angular approximations and asymptotic expansions.

Given a self-adjoint operator A on $\mathbb{R}^d$ ($d \geq 1$) or the corresponding symmetric matrix (we do not notationally distinguish them), we denote its spectrum by $\mathrm{Spec}(A) \subset \mathbb{R}$. We consider a family (A_μ) of self-adjoint operators on $\mathbb{R}^d$, which we assume is defined for $|\mu| \leq \mu_0$ for some $\mu_0 > 0$. We do not assume any regularity with respect to the parameter μ and indeed the results in this Paragraph 2.3 will be applied to situations where regularity in μ cannot be ensured; typically the canonical transformations $\mathcal{C}_\varepsilon$ constructed above are not regular with respect to ε. We assume however that A_μ is μ-close to A_0; more precisely we denote by $C > 0$ a constant such that
$$\|A_\mu - A_0\| \leq C|\mu|.$$
So C can be taken to be the C^1-norm of (A_μ) in case this family happens to be of class C^1. Here and below we use the operator norm for the matrices, but since we are dealing with finite-dimensional problems this is not essential.

Let $\mathrm{Spec}(A_0) = \{\lambda_1, \ldots, \lambda_k\}$ ($\lambda_i \neq \lambda_j$ for $i \neq j$, $k \leq d$) be the spectrum of A_0 and m_j be the multiplicity of λ_j so that $\sum_j m_j = d$. The first result is the

LEMMA (Spectral continuity of eigenvalues). *Denote by $\|A\|$ the operator norm of A. Then the spectrum of A_μ varies continuously with μ and in fact $\mathrm{Spec}(A_\mu)$ is contained in a $\|A_\mu - A_0\|$-neighborhood of $\mathrm{Spec}(A_0)$:*
$$dist\bigl(\mathrm{Spec}(A_\mu), \mathrm{Spec}(A_0)\bigr) \leq \|A_\mu - A_0\| \leq C|\mu|.$$

Moreover if the family (A_μ) is analytic (with respect to μ), there exists a labeling of the eigenvalues which makes $\mathrm{Spec}(A_\mu)$ analytic.

PROOF. The first part of the lemma is an immediate consequence of the min-max principle (see *e.g.* [RS], vol. 4, §XIII.1). It is thus quite specific to self-adjoint operators, whose eigenvalues vary in a Lipschitz way, and indeed with a Lipschitz constant equal to 1. In the terminology of numerical analysis, they are "well-conditioned". The second part of the lemma, which we have stated slightly informally for the sake of brevity, is classical; for details and much more, see *e.g. loc. cit.* Chapter XII. □

Let us say a word about the situation when A_μ is *not* assumed to be self-adjoint. Then there emerges from λ_j at most m_j eigenvalues, which can be labeled just as in the self-adjoint case. But if (A_μ) is analytic, these eigenvalues bifurcating from λ_j are given by Puiseux series in general, more precisely they are analytic with respect to μ^{1/m_j} near $\mu = 0$. Correspondingly, in the C^1 case, one can only assert that $\mathrm{Spec}(A_\mu)$ lies in a neighbourhood of $\mathrm{Spec}(A_0)$ of size $O(\|A_\mu - A_0\|^{1/m})$ where $m = \sup_j m_j$ and the implied constant is not universal (in particular is not 1 as in the self-adjoint case).

We now turn to the stability properties of the eigenvectors. For any $j = 1, \ldots, k$, let d_j be the distance from λ_j to the rest of the spectrum of A_0: $d_j = \inf_{i \neq j} |\lambda_i - \lambda_j|$. For μ small enough, as mentioned above the spectrum of A_μ contains exactly m_j eigenvalues counted *with* multiplicities inside the disc $D_j = D(\lambda_j, d_j/2)$ centered at λ_j with radius $d_j/2$ (the spectrum is real of course, but we introduce D_j in the complex plane for future use). Let $\Pi_\mu^{(j)}$ denote the eigenprojector corresponding to these eigenvalues of A_μ contained in D_j; it is thus an orthogonal projector whose range has dimension m_j. Because of the preceding lemma it is easy to quantify the bound on μ: in fact we define μ_j by requiring that for $|\mu| < \mu_j$ one has $\|A_\mu - A_0\| < d_j/4$. This is of course provided A_μ is defined in this range; if not just define $\mu_j = \mu_0$. Using the definition of C one gets an obvious bound from below for μ_j, namely $\mu_j \geq d_j/(4C)$. The projector $\Pi_\mu^{(j)}$ is well-defined for $|\mu| < \mu_j$. It satisfies:

LEMMA (Spectral continuity of eigenprojectors). *For any $j \in \{1, \ldots, k\}$ and for $|\mu| < \mu_j$ with μ_j defined as above (in particular $\mu_j \geq d_j/(4C)$), one has the estimate:*
$$\|\Pi_\mu^{(j)} - \Pi_0^{(j)}\| < c_j |\mu|,$$
where $c_j = 8C/d_j$. If the family (A_μ) is analytic, the orthogonal projector $\Pi_\mu^{(j)}$ is analytic too, for $|\mu| < \mu_j$.

PROOF. It is based on functional calculus, for which we refer to [RS], vol. 4, §XII.1,2, as well as for background material on the regular perturbation theory of self-adjoint operators. One can express the projectors as

(15) $$\Pi_\mu^{(j)} = \frac{1}{2\pi i} \oint_{C_j} (\lambda - A_\mu)^{-1} d\lambda,$$

where $C_j = \partial D_j$ (anticlockwise orientation). This formula implies the existence of c_j, but not yet the estimate given in the statement. It also implies that $\Pi_\mu^{(j)}$ is as regular as the family (A_μ), so also the second part of the lemma, namely the analyticity of the projectors in the analytic case.

Now to get the bound on c_j, we simply compute $\Pi_\mu^{(j)} - \Pi_0^{(j)}$ by taking the difference of the r.h.s. of (15) for A_μ and A_0; one gets inside the integral the quantity
$$(\lambda - A_\mu)^{-1} - (\lambda - A_0)^{-1} = (\lambda - A_\mu)^{-1}(A_\mu - A_0)(\lambda - A_0)^{-1},$$
to be integrated over C_j. Now for any self-adjoint operator A and any λ in its resolvent set one has $\|(\lambda - A)^{-1}\| = \bigl(\mathrm{dist}(\lambda, \mathrm{Spec}(A))\bigr)^{-1}$. This implies that $\|(\lambda - A_\mu)^{-1}\| < 4/d_j$ for $\lambda \in C_j$ and $|\mu| < \mu_j$; note that this estimate makes full use of the preceding lemma. One finishes the proof of the lemma by obvious bookkeeping. □

Let us say again a word about the case of arbitrary matrices. Then the lemma is true *verbatim* except for the estimates on μ_j and c_j and the following important difference. The projectors $\Pi_\mu^{(j)}$ are *defined* by the r.h.s. of (15) for μ small enough, without a universal explicit estimate, for the reasons explained after the lemma on eigenvalues. They are not orthogonal in general and their range is a direct sum of characteristic spaces which we recall are defined as follows (think of the Jordan normal form for a matrix): given an operator A on a vector space of dimension d and $\lambda \in \mathbb{C}$, the associated characteristic space V_λ is the kernel of $(A - \lambda)^d$; λ is called characteristic if V_λ is not reduced to 0.

2.3.2. We also need to be able to record the behaviour of the spectral elements of the splitting matrix under canonical transformations. The variance of the matrix itself is expressed in Proposition 1.5.4 and in order to use it we need another lemma in linear algebra, which we first state and prove in a non-perturbative version. We remark that the properties it expresses are again specific to self-adjoint operators.

Let A be a symmetric operator on $\mathbb{R}^d$ ($d \geq 1$), and let B and C be two operators such that C is invertible and the composition BAC is symmetric (B and C are *not* assumed to be symmetric). Suppose that A possesses n eigenvalues (counting with multiplicities) in the interval $]-\delta, \delta[$ for some $\delta > 0$. Let Π be the corresponding spectral projector for A and V the range of Π. It is an n-dimensional subspace of $\mathbb{R}^d$ which is stable under A and such that the restriction of A to V has norm $< \delta$. In this situation, we get the following information on the spectral elements of the operator BAC:

LEMMA. *Given the data as above, the following holds:*
i) The operator BAC has at least n eigenvalues (counted with multiplicity) of modulus $< \delta \|B\| \cdot \|C^{-1}\|$;
ii) There is an n-dimensional spectral projector Π' of the operator BAC, corresponding to eigenvalues as above and close to the spectral projector Π of A in the sense that: $\|\Pi' - \Pi\| \leq \|B - 1\| \cdot \|C^{-1} - 1\|$.

PROOF. Consider the n-dimensional subspace $W = C^{-1}V$. For any $w \in W$, one has $\|BAC(w)\| < \delta \|BC(w)\|$ because $C(w)$ is an element of V and over that subspace the norm of A is $< \delta$. This proves i) by the min-max principle and ii) follows in the same fashion. □

We will have to use also a perturbative version of this lemma. Namely let B_ε and C_ε be two families of operators defined for ε small enough, such that $B_0 = C_0 = \mathrm{Id}$ and $B_\varepsilon A C_\varepsilon$ is symmetric; here A again denotes a fixed self-adjoint operator. We require again that $B_\varepsilon = B_0 + O(\varepsilon)$ and the analogous condition for C_ε. The assumptions on A are as in the lemma and one gets the conclusion of the lemma for ε small enough, with of course $\|B_\varepsilon\| \cdot \|C_\varepsilon^{-1}\| = 1 + O(\varepsilon)$ and $\|B_\varepsilon - 1\| \cdot \|C_\varepsilon^{-1} - 1\| = O(\varepsilon)$. This continues to hold if the operator A is replaced by a family A_μ, which does not affect the reasoning (δ becomes a function of μ).

2.3.3. We now look back at the formalism developed in Chapter 1 and first prove strong smoothness properties for the various angular notions because this is good to have in store and the proof is now essentially immediate. Recall that the Lagrangian-Grassmannian Λ of $\mathbb{R}^\ell$ is a smooth manifold with an atlas $(\Lambda_V)_{V \in \Lambda}$ (*cf.* §1.3.1) which actually defines a *real analytic* structure on Λ. In turn the angular notions describe in some sense various atlasses on Λ^2. Although that would be feasible we will not explore the whole of Λ^2 but confine ourselves to a large open set for simplicity. Define $\mathcal{O} \subset \Lambda^2$ to be the open set of pairs (L_1, L_2) with L_2 transverse to $\mathrm{Orth}_\Pi(L_1)$ (Π is the usual scalar product on $\mathbb{R}^{2\ell}$); so $\mathcal{O}$ is a—large— neighbourhood of the diagonal. Given two transverse Lagrangian subspaces (H, V), let $\mathcal{O}_{HV}$ be the open subset of $\mathcal{S}(H)^2$ (*cf.* §1.3.1) formed by the pairs (γ_1, γ_2) such that, if L_i ($i = 1, 2$) is the Lagrangian space with coordinate γ_i relative to (H, V), the pair (L_1, L_2) lies in $\mathcal{O}$. Then looking back at §§1.3–1.4 we have:

PROPOSITION. *The maps*

$$\begin{cases} \mathcal{O}_{HV} \to U(n, \mathbb{C}) \\ (\gamma_1, \gamma_2) \mapsto \varphi_{L_1, L_2} \end{cases} \quad and \quad \begin{cases} \mathcal{O} \to U(n, \mathbb{C}) \\ (L_1, L_2) \mapsto \varphi_{L_1, L_2} \end{cases}$$

are real analytic. The measure of the Euclidean angle of (L_1, L_2) *is also real analytic over* $\mathcal{O}$.

PROOF. We may assume that $H = H_0$, that $V = V_0$ and furthermore that L_1 is fixed and coincides with H_0. By the spectral lemma on the eigenvalues (in §2.3.1) we see that the spectrum of γ is analytic and by the definition of φ, we find that this implies that $\gamma \mapsto \varphi(\gamma)$ is analytic (*cf.* §1.4.1). Since γ defines the analytic topology, we conclude that both maps are analytic and that the measure of the angle is no other but the spectrum of φ. □

One would have to be more cautious in order to deal with the determination of the angle, *i.e.* with the eigenspaces of φ (or γ since they coincide). Indeed we would have to modify the definition a bit in order to get good smoothness properties, but we will not go into this here.

2.3.4. This paragraph is in some sense a sequel to §§1.5–1.7. Starting from the general symplectic setting of Chapter 1, one can contemplate several simplifications or specializations and it is good to keep in mind that the situation of "classical perturbation theory", as used above in §§2.1–2.2, is special in more than one way. Here we list a few items in order to bridge the gap between geometry and "computations in coordinates". In doing so we set up the framework which we use in further paragraphs and in the next chapter; it is essentially that of §1.9 which on the whole appears as a rather particular case of the situations described in Chapter 1. It can be used independently if one wishes to forget about more global and geometric considerations.

First we specialize from an arbitrary symplectic manifold to a cotangent bundle T^*V. We recall from §1.7 that any such bundle is endowed with a natural vertical distribution $\mathcal{V}$ given by the fibers of the projection $T^*V \to V$, but not with a horizontal one (except along the zero section). Such a distribution $\mathcal{H}$ is associated with the further data of a connection on V. Then one can look at a Lagrangian submanifold $\mathcal{L}$ and ask whether it is a graph over $\mathcal{H}$, in which case it is defined by an embedding $\beta : V \to T^*V$, where β is a closed one-form on V. If β is furthermore exact, $\beta = dS$ where the real valued function S on V is a generating function for $\mathcal{L}$. The relevant formulas in this situation appear in §§1.7.1–1.7.2.

We will be interested in regular pairs of Lagrangian submanifolds $(\mathcal{L}_1, \mathcal{L}_2)$ which are close to each other in the sense that, for any $x \in V$, the pair $(L_1, L_2) = (T_x\mathcal{L}_1, T_x\mathcal{L}_2)$ lies in the open set $\mathcal{O} \in \Lambda^2$ defined above in §2.3.3. Given the reference distributions $(\mathcal{H}, \mathcal{V})$, the associated coordinates (γ_1, γ_2) of (L_1, L_2) with respect to the pair $(H, V) = (T_x\mathcal{H}, T_x\mathcal{V})$ will lie in $\mathcal{O}_{H,V}$; this assumes, as we do, that $\mathcal{H}$ and $\mathcal{V}$ are orthogonal for a given orthogonal structure on T^*V.

Next assume that we are in a perturbative situation, in the sense that we have smooth families $\mathcal{L}_1(\delta)$ and $\mathcal{L}_2(\delta)$ defined for δ small enough and with $\mathcal{L}_1(0) = \mathcal{L}_2(0) = \mathcal{L}$. Note that we do not assume as yet that $\mathcal{L}$ is close to $\mathcal{H}$. Because we will be interested in local properties, and in order to avoid cumbersome assumptions and notation, we will work in a linear setting, using the notation of §§1.3–1.5. All the spaces of linear maps are endowed with their natural operator norms, which are all denoted by the symbol $\|\ \|$ (since everything is finite-dimensional all the possible norms are equivalent anyway).

Consider first the case where $L_1(\delta) = H$ is horizontal and $L_2(\delta) = L(\delta) \in \Lambda_H$ with coordinate $\gamma(\delta)$ ($\gamma(0) = 0$). One cannot of course simplify the step consisting in finding the spectral elements of γ, so we also give ourselves an eigenbasis for γ, with eigenvalues $\alpha_j(\delta)$, $j = 1, \ldots, k$. One can assert, using the spectral lemma for eigenvalues, that $\alpha_j = O(\delta)$ so that (*cf.* §1.4.1)

$$(16) \qquad \rho_j = \exp(i\theta_j) = \frac{1 + i\alpha_j}{|1 + i\alpha_j|} = \exp(i\alpha_j) + o(\delta).$$

The eigenvalues of γ thus give a first-order approximation for the measure of the Euclidean angle (with the horizontal). Indeed the Euclidean angle with respect to the horizontal plane and the coordinate carry the same information near the horizontal plane (essentially because θ and $\tan(\theta)$ are close for θ small).

In the above, one of the planes was horizontal; in the general case of a pair (L_1, L_2) of nearby Lagrangian planes, the situation is different according to whether one is interested in the measure of the Euclidean angle or with the difference of the coordinates *i.e.* the splitting form or splitting matrix (*cf.* §1.5). Roughly speaking the first is geometrically more significant and the second is easier to compute. The case of the Euclidean angle is conceptually quite easy: given (L_1, L_2) (depending on δ which we sometimes drop from the notation) there is a smooth family of unitary maps $\varphi_{H_0L_1}^{-1}$ such that setting $L = \varphi_{H_0L_1}^{-1}(L_2)$ we are back to the case above, *i.e.* the Euclidean angle of the pair (L_1, L_2) is the same as the angle of (H_0, L).

For practical purposes it is easier to work with the splitting matrix and the formulas which appear in §1.5. Let $\varphi_{H_0L_1}^{-1}$ be as above. Let $\mathbf{\Gamma}_i$, $i = 1, 2$, be the coordinate of L_i, $\mathbf{S} = \mathbf{\Gamma}_2 - \mathbf{\Gamma}_1$ be the splitting matrix. Let $\mathbf{\Gamma}$ denote the coordinate of L. Since $\varphi_{H_0L_1}$ maps (H, L) to (L_1, L_2), the splitting matrix after performing $\varphi_{H_0L_1}^{-1}$ coincides with $\mathbf{\Gamma}$. So $\mathbf{\Gamma} = \mathbf{S}'$ as given by the formula in Proposition 1.5.4. The matrix $\mathbf{P}$ (*cf.* §1.5.1) corresponding to φ cannot be computed in general. All that one can say is that it remains finite as long as L_1 and L_2 stay away from V.

Finally, assume that the Lagrangian planes L_1 and L_2 are close to each other *and* close to the horizontal plane H. Then $\varphi_{H_0L_1}$ is close to the identity, say $\varphi_{H_0L_1} = \mathrm{Id} + O(\varepsilon)$ with ε small (tending to 0 with δ). Typically in the case of an exponentially small splitting, L_1 and L_2 are the stable and unstable invariant manifolds attached to a torus. In that case δ will be typically exponentially small with respect to ε. Using again Proposition 1.5.4 we get the estimate: $\mathbf{\Gamma} = (1 + O(\varepsilon))\mathbf{S}(1 + O(\varepsilon))$.

We also recall that Proposition 1.5.4 gives the variance of the splitting matrix under a canonical transformation. This is just a different viewpoint, but computations are identical. In particular, for a transformation which is ε-close to identity, one gets the same estimate as above, namely that the splitting matrices before and after the transformation are connected by: $\mathbf{S}' = (1 + O(\varepsilon))\mathbf{S}(1 + O(\varepsilon))$. This suggests that Lemma 2.3.2 applies, as will be done below.

The above did not depend on the global nature of the underlying manifold (the "configuration space") V. As a final specialization take $V = \mathbb{T}^\ell$. Then T^*V is trivial and there is a natural trivialization: $T^*\mathbb{T}^\ell \simeq \{(\phi, I) \in \mathbb{T}^\ell \times \mathbb{R}^\ell\}$. Note that it is more consistent to list the coordinates as (ϕ, I) (the base and then the fibre), but we will sometimes write (I, ϕ) all the same. Now we have natural horizontal ($I =$ cst) and vertical ($\phi =$ cst) distributions. In this very special (from the geometric viewpoint) case, we can also ignore the difference between linear and global notions, so forget about calligraphic letters. Lastly we usually make no notational difference between ϕ as an element of the torus $\mathbb{T}^\ell$ or its universal cover $\mathbb{R}^\ell$.

Referring again to the beginning of this paragraph (or to §1.7), we will deal with Lagrangian manifolds L which are graphs of one-forms β on the torus: we write $L = \mathcal{G}r(\beta)$. As mentioned in §1.7.1, the defect of exactness is measured by an ℓ-vector. Concretely one can write $\beta = dS(\phi)$ with $\phi \in \mathbb{R}^\ell$ and there exists $\delta \in \mathbb{R}^\ell$ such that $S(\phi) - \delta\phi$ is a function on $\mathbb{T}^\ell$. So we always have a generating function $S(\phi)$ for β which up to a linear factor is defined on $\mathbb{T}^\ell$. Formally we can always consider it as defined on $\mathbb{R}^\ell$ or over an open fundamental domain, e.g. $]-\pi,\pi[^\ell$. These details will not play any role anyway, partly because we are dealing with the *homo*clinic splitting problem; we simply write $\beta = dS(\phi)$. Then the coordinate γ is given by the matrix $\mathbf{\Gamma} = (\mathbf{\Gamma}_{ij}) = (\partial^2 S / \partial\phi_i \partial\phi_j)$ (see §1.7.2 for the more general setting). Finally, given two intersecting Lagrangian manifolds L_1 and L_2, the splitting matrix at an intersection point is given by the difference $\mathbf{S} = \mathbf{\Gamma}_2 - \mathbf{\Gamma}_1$. We also note that this is the setting which will be used in Chapter 3 as well as below.

2.4. General exponential estimates for the splitting matrix

We return to the situation of Paragraphs 2.1 and 2.2 and prove a provisional result asserting that, *if* the necessary invariant objects exist, *then* the splitting matrix has a certain form and in particular some of its eigenvalues are exponentially small. The existence of the invariant objects will be discussed under some conditions in §2.5, using KAM theory and the results of Chapter 1. The results in this section apply in a very general (but perturbative) setting and the need for the existence proofs taken up in §2.5 arises from this generality. More precisely the existence of the necessary objects is easier to study if one restricts to *simple* resonances; it becomes obvious for, say, Hamiltonian (∗) with F independent of the action variables and even in the angles, which covers the cases hitherto studied in the literature. We also stress the important fact that the present paragraph does *not* make direct use of any analyticity conditions, which however were crucial in order to derive the normal form (3) with an exponentially small remainder.

2.4.1. We start with some notation and assumptions. It is actually a bit difficult to stick to completely uniform assumptions for the numerous possible statements, so we ask the reader to be forgiving about the slight variations which are bound to occur and will be recorded all along. We hope that the general ideas will be clear and that the reader might want to state and prove variants of his own.

Let $L_\mu^\pm$ be two families of Lagrangian submanifolds in $T^*\mathbb{T}^\ell$. We write $L_\mu^\pm$ rather than e.g. $L_1(\mu)$ and $L_2(\mu)$ simply because these will of course eventually be invariant manifolds, but right now this is just a piece of notation. We assume that $L_\mu^\pm$ is defined for μ small enough and that $L_\mu^\pm = \mathcal{G}r(dS_\mu^\pm)$. Let us add a minor global warning for this and the next subsections §§2.4–2.5. Some of the Lagrangian manifolds we will consider are *not* necessarily exact, which means that S_μ does not necessarily live on the torus $\mathbb{T}^\ell$. In other words we should consider ϕ as belonging to $\mathbb{R}^\ell$ and indeed restrict to bounded domains $V^\pm$ for $S_\mu^\pm$. Such a function can be uniquely decomposed as $S_\mu = S'_\mu + \delta \cdot \phi$, where the added linear form encodes the translation just as in §1.10. Since the splitting matrix involves the *second* derivative of the function S_μ (see (18) below), this translation does not enter in a significant way here. This is why we will often omit from the notation the distinction between functions on the torus $\mathbb{T}^\ell$ and its covering space $\mathbb{R}^\ell$, a distinction which the reader can easily mentally restore. Note also that we do not require any regularity in μ

but assume that $S_\mu^\pm$ is a C^2 real-valued function on $V^\pm$ and that $S_\mu^\pm - S_0^\pm = O(\mu)$ in the C^2 topology of functions on $V^\pm$. Of course we can vary $V^\pm$, which amounts to looking at the compact-open C^2 topology. The only precaution to be taken is that of course the implied constant in the $O(\mu)$ estimate above depends on the domains $V^\pm$. More generally the non-compactness of the invariant manifolds leads to slight problems in the formulation of the results but not to serious mathematical difficulties in what follows. Even the graph requirement here can be removed, as will be mentioned at appropriate places.

Assume moreover that L_μ^+ and L_μ^- intersect at a (not necessarily isolated) point p_μ. We have a splitting matrix $\mathbf{S}_\mu$ at the point p_μ, which is given as in §2.3.4 above by:

$$\mathbf{S}_\mu(p_\mu) = \frac{\partial^2}{\partial \phi^2}(S_\mu^+ - S_\mu^-)(p_\mu). \tag{17}$$

Here the evaluation at p_μ means of course that one looks at the angular coordinate of p_μ.

2.4.2. Consider again the decompositions (3) and (14), taking care of the fact that the variables in (14) have been rescaled according to (8) (and primes have been dropped in (14)). Summarizing what was done so far, starting from (1) we have performed a canonical transformation $\mathcal{C}_\varepsilon$ which is ε-close to identity and leads to (3) where R_ε is exponentially small, as indicated in Proposition 2.1.4. Writing H for $H \circ \mathcal{C}_\varepsilon$, we thus have $H = H_r + R_\varepsilon$ and moreover $H_r = H_r^0 + O(\sqrt{\varepsilon})$ as in (14), where this last decomposition is in terms of the variables rescaled as in (8).

Let $\mu = \varepsilon^{-1}\|R_\varepsilon\|$ and write $H = H_r + \varepsilon\mu\tilde{R}_\varepsilon$ where $\tilde{R}_\varepsilon$ has norm 1. Consider μ as an independent perturbation parameter so that we have a two-parameter family of Hamiltonian systems (a trick initiated by Poincaré). Of course we are really interested in the value $\mu = \mu(\varepsilon)$ dictated by the definition of μ and this value is exponentially small with respect to ε. We have added the factor ε in front in the definition of μ so that in the rescaled variables of (14) we can write $H = H_r + O(\mu) = H_r^0 + O(\sqrt{\varepsilon}) + O(\mu)$. We remark that not much depends on the fact that we are eventually interested in exponentially small values of μ (with respect to ε) and most results can be stated in terms of μ and then specialized to much larger values (with identical proofs). For instance in this paragraph we could actually accomodate polynomial values of μ, say up to $\mu = \varepsilon^2$.

By now the strategy should be clear. We start from the generalized pendulum with m degrees of freedom defined by the Hamiltonian $P(I_2, \phi_2) = \frac{1}{2}BI_2^2 + g(\phi_2)$ (cf. §2.2). We then suspend it into H_r^0, which is in principle straightforward. The next step consists in "turning on" the perturbation parameter ε keeping $\mu = 0$, which is possible as they are now regarded as independent parameters. We are moving from H_r^0 to H_r, that is incorporating the $O(\sqrt{\varepsilon})$ remainder in (14) and this is again ϕ_1-independent. At that point one can use the results of §1.9 in order to straighten the normally hyperbolic manifolds in the resonant system governed by H_r (see §2.5.1). This we will do in order to be able to apply the KAM results contained in [Gr]. Next we have to let μ take off from zero and this is yet another story, involving in particular KAM techniques, which we discuss briefly in §2.5.2. But it seems useful to distinguish between assumptions which ensure the *existence* of the objects and those which govern their *properties*, granted that they exist; this paragraph is concerned primarily with the *second* kind. In the final step we must remember that in the above we were working in the normalizing variables and that we have to return to the original variables in (1) *via* the scaling (8) and more seriously the near-identity canonical transformation $\mathcal{C}_\varepsilon$.

2.4.3. Let us review the geometric objects we have at our disposal or need to locate, according to §2.2.2. Start with H_r^0 which itself decomposes as the generalized pendulum $P(I_2, \phi_2)$ and a part depending on I_1 only. This holds after the change of variables (12) which is however of no concern in this paragraph. Assume B is strictly positive as in §2.2.2 (with obvious adaptations if it is strictly negative) and let $x_0 \in T^*\mathbb{T}^m$ be the hyperbolic fixed point of X_P which is discussed there (recall that X_P denotes the vector field of $T^*\mathbb{T}^m$ generated by the Hamiltonian function P). Let $W^\pm(x_0)$ denote the m-dimensional invariant manifolds of x_0. For simplicity we assume hereunder that these are graphs on the domains we are interested in, namely along relatively compact parts of the homoclinic trajectory we will be interested in (see §1.10 and below for more details). This assumption is actually superfluous because in fact we use symplectic geometry only near the homoclinic points where we eventually compute the splitting matrices, but it makes it easier to write down explicit expressions. So here we write:

$$W^\pm(x_0) = \mathcal{G}r(dS_r^\pm) \subset T^*\mathbb{T}^m, \quad \text{with} \quad S_r^\pm : \mathbb{T}^m \to \mathbb{R},$$

where the functions $S_r^\pm(\phi_2)$ are only supposed to be defined on the domains we are interested in.

Coming to the homoclinic splitting, we clearly need a homoclinic trajectory in order to get started. The existence of such trajectories can in principle be shown by variational methods (see §2.5.4) in this non-perturbative situation. Here we just *assume* that $W^+(x_0)$ and $W^-(x_0)$ do intersect. More precisely their intersection is necessarily a union of trajectories of the Hamiltonian P; we denote by γ one of these, and assume that $W^+(x_0)$ and $W^-(x_0)$ intersect *transversely* along γ in the energy level. This is a generic assumption. We next pick a point p on this homoclinic trajectory:

$$p \in \gamma \subset W^+(x_0) \cap W^-(x_0).$$

As in §1.10, we consider two relatively compact neighborhoods $D^\pm$ in $\mathbb{T}^m$ of the projections of the two half-trajectories from p to x_0. Below the discussion will be restricted to these sets and we assume that the invariant manifolds $W^\pm(x_0)$ are graphs over $D^\pm$; this assumption is inessential but facilitates the exposition.

In order to eventually define nice splitting matrices we will now perform a section Σ_r (*cf.* §§1.1.4 and 1.5.5) transverse to γ at p inside the energy surface of γ (which is the same as that of the fixed point x_0). For reasons which will appear in the next paragraph, it is convenient to choose a section of the form

$$\Sigma_r = \tilde{\Sigma}_r \cap \{P = P(x_0)\},$$

where $\tilde{\Sigma}_r$ is a hypersurface of $T^*\mathbb{T}^m$ which contains p, is transverse to X_P and sits over $D^+ \cap D^-$.

Then we consider the splitting matrix at p of the traces of the invariant manifolds $W^+(x_0)$ and $W^-(x_0)$ on Σ_r. We denote this matrix by $M_0^\perp$ because it represents the linear approximation of the splitting of (1) transverse to the m-fold resonance surface we are looking at. For the same reason we call it the *transverse Poincaré-Melnikov splitting matrix* or *transverse linear splitting matrix*. It is a square symmetric matrix of size $m-1$; in particular it appears only for resonances of multiplicity $m > 1$. Formula (VΣ) in §1.5.5 shows that it transforms like a quadratic form upon changing the section, which by the min-max principle (as embodied in Lemma 2.3.2) guarantees that the choice of a section is innocuous in what follows.

The hypothesis that $W^+(x_0)$ and $W^-(x_0)$ intersect transversely along γ is equivalent to the nondegeneracy of the matrix $M_0^\perp$. Concretely one can simply compute $\mathbf{S}_0^\perp(p) = \partial^2/\partial\phi_2^2(S_r^+ - S_r^-)(p)$, and requiring that the transverse splitting be nondegenerate is the same as requiring that $\mathbf{S}_0^\perp(p)$ have only one 0 eigenvalue (corresponding to the eigendirection tangent to γ). Note that we are now in the setting of §§1.10.1–1.10.2 ($m=1$ or $m > 1$).

Having described the situation for X_P in $T^*\mathbb{T}^m$, we could easily lift everything to $T^*\mathbb{T}^\ell$ for $X_{H_r^0}$. But we prefer to pass from H_r^0 to H_r at the same time, *i.e.* to introduce the parameter ε.

2.4.4. Since the Hamiltonian function H_r does not depend on ϕ_1, the conjugate variable I_1 provides n prime integrals of X_{H_r} (this corresponds to the case $\mu = 0$ in §1.9). Let us fix an arbitrary value of I_1: $I_1^0 \in \mathbb{R}^n$.

By restricting to $\{I_1 = I_1^0\}$ and projecting to $T^*\mathbb{T}^m$, the vector field X_{H_r} induces a vector field of $T^*\mathbb{T}^m$ which is $\sqrt{\varepsilon}$-close to X_P. The hyperbolic fixed point x_0 of X_P is perturbed into a hyperbolic fixed point $x_\varepsilon(I_1^0)$, and its invariant manifolds $W^\pm(x_0)$ deform into invariant manifolds $W^\pm(x_\varepsilon(I_1^0))$. Moreover, the transversality assumption at the homoclinic point p ensures the existence of a homoclinic point $p_\varepsilon(I_1^0) \in \Sigma_r$ close to it. The *transverse splitting matrix* $M_\varepsilon^\perp$ at $p_\varepsilon(I_1^0)$ in Σ_r is a $O(\sqrt{\varepsilon})$-perturbation of $M_0^\perp$. We denote by $S_{r,\varepsilon}^\pm(\phi_2; I_1^0)$ the new generating functions. They depend on I_1^0 as a parameter, and $S_{r,0}^\pm(\phi_2; I_1^0) = S_r^\pm$.

All this can be interpreted for X_{H_r} by lifting the objects to $\{I_1 = I_1^0\} \subset T^*\mathbb{T}^\ell$: there corresponds to $x_\varepsilon(I_1^0)$ an invariant n-torus $\mathcal{T}_\varepsilon(I_1^0)$ spanned by ϕ_1 (as claimed in Proposition 2.2.2), which has ℓ-dimensional invariant manifolds $\mathcal{W}_\varepsilon^\pm(I_1^0)$. These manifolds are given by:

$$\mathcal{W}_\varepsilon^\pm(I_1^0) = \mathcal{G}r(dS_\varepsilon^\pm) \quad \text{with} \quad S_\varepsilon^\pm = S_\varepsilon^\pm(\phi; I_1^0) = I_1^0 \cdot \phi_1 + S_{r,\varepsilon}^\pm(\phi_2; I_1^0).$$

The functions $S_\varepsilon^\pm$ are defined over the domains $V^\pm = \mathbb{T}^n \times D^\pm$. Here it may be necessary to perform a restriction of the domain in the I_1-variables, so that $V^\pm$ are independent of I_1. But in the end we will be interested only in a neighborhood of $I_1 = 0$, after turning on the μ-perturbation. Note also that the invariant manifolds may not be exact Lagrangian at this point so that the functions $S_{r,\varepsilon}^\pm$ really live *a priori* on $\mathbb{R}^\ell$ and possibly include a linear term in ϕ_2. In accordance to the remark at the end of §2.3 we do not record this explicitly in the notation.

In particular, for $\varepsilon = 0$, we obtain $S_0^\pm(\phi; I_1^0) = I_1^0 \cdot \phi_1 + S_r^\pm(\phi_2)$. Note that here the change of variables (12) which was used to eliminate the mixed terms in the quadratic form (10) simply introduces an extra linear factor $B^{-1}CI_1^0 \cdot \phi_2$ in the generating functions $S_0^\pm(\phi)$ and this factor does not affect the argument below. Note also that the crux of the matter in showing general exponential estimates eventually consists in the (trivial) fact that $S_\varepsilon^\pm$ is linear with respect to ϕ_1.

In order to compute a splitting matrix, we now have to fix a section in $T^*\mathbb{T}^\ell$ and a particular homoclinic point. The torus $\mathcal{T}_\varepsilon(I_1^0)$ comes indeed with a $(n+1)$-dimensional homoclinic manifold $\mathcal{W}_\varepsilon^+(I_1^0) \cap \mathcal{W}_\varepsilon^-(I_1^0)$. As for the section, we choose

$$\Sigma_\varepsilon(I_1^0) = \tilde{\Sigma} \cap \{H_r = H_r(\mathcal{T}_\varepsilon(I_1^0))\},$$

where $\tilde{\Sigma}$ denotes the hypersurface of $T^*\mathbb{T}^\ell$ obtained by taking the product of $\tilde{\Sigma}_r$ with $T^*\mathbb{T}^n$. The dependence of $\Sigma_\varepsilon(I_1^0)$ thus occurs *via* the choice of I_1^0; for any I_1^0 the section $\Sigma_\varepsilon(I_1^0)$ sits over $V^\pm$.

This way, the intersection

$$\mathcal{W}_\varepsilon^+(I_1^0) \cap \mathcal{W}_\varepsilon^-(I_1^0) \cap \Sigma_\varepsilon(I_1^0) = \mathfrak{T}_\varepsilon(I_1^0)$$

is a homoclinic n-torus biasymptotic to $\mathcal{T}_\varepsilon(I_1^0)$ for the system governed by H_r.

The torus $\mathfrak{T}_\varepsilon(I_1^0)$ is canonically parametrized by ϕ_1. Let us denote by $p_\varepsilon(\phi_1; I_1^0)$ the point of $\mathfrak{T}_\varepsilon(I_1^0)$ with parameter ϕ_1. We observe that the second derivatives $\partial^2 S_\varepsilon^\pm/\partial\phi_1^2$ and $\partial^2 S_\varepsilon^\pm/\partial\phi_1\partial\phi_2$ vanish at this (fixed but arbitrary) point $p_\varepsilon(\phi_1; I_1^0)$. This implies that the splitting matrix M_ε at this point in the section $\Sigma_\varepsilon(I_1^0)$ takes the form:

$$(18) \qquad M_\varepsilon = \begin{pmatrix} 0 & 0 \\ 0 & M_\varepsilon^\perp \end{pmatrix}.$$

It is an $(\ell-1)$-symmetric matrix, which has a zero upper-left block of size n, zero off-diagonal blocks and the transverse splitting matrix $M_\varepsilon^\perp$ of size $m-1$ in the lower right corner.

In particular, in the case of H_r^0 (*i.e.* when $\varepsilon = 0$),

$$M_0 = \begin{pmatrix} 0 & 0 \\ 0 & M_0^\perp \end{pmatrix}.$$

We remark that the nondegeneracy of the transverse Poincaré-Melnikov matrix $M_0^\perp$ has been used to ensure the *existence* of the necessary objects, in particular the homoclinic tori $\mathfrak{T}_\varepsilon(I_1^0)$. But if these are known to exist by some other considerations, one can still assert that the splitting matrix looks as in (18).

2.4.5. We digress shortly in order to discuss an analytic point about domains of validity. Everything above takes place in the normalized coordinates defined by Proposition 2.1.4, which after the rescaling (8) give rise to (14). We need that the objects of interest be contained in the domain of the normalizing transformation $\mathcal{C}_\varepsilon$. As far as the variables I_1 are concerned, we are eventually interested only in a neighborhood of the origin of size $O(\sqrt{\varepsilon})$ in the variables used in (14), which we call the rescaled normalized variables. This corresponds to the fact that the shift of the torus we started with (for $\varepsilon = 0$ in (1)) is on the order of ε in the original variables of (1). The real question concerns the resonant variables I_2 and what we need to ensure is again that the homoclinic point p together with the two half-trajectories from p to the fixed point x_0 be contained in the domain of existence of the normalized variables. But by accepting to possibly decrease the width w of the exponentially small remainder R_ε in Proposition 2.1.4, one can make the size ρ of the domain arbitrarily large after rescaling (see under (14)). In more physical terms this can be expressed by saying that one can normalize to an exponentially large order over a strip surrounding the resonance which is $O(\sqrt{\varepsilon})$ wide (the "resonance width") with an arbitrarily large coefficient in front and keeping the exponent (but not the width) fixed. This ensures the coherence of the strategy. We will return to this point briefly and in more general terms in the conclusion of the closing section §3.7.4.

2.4.6. We now have to restore "full non-integrability", that is perturb the resonant part H_r into the original system H, although still in the rescaled normalized coordinates. At this point, we would like to focus attention on the torus in the family $\mathcal{T}_\varepsilon(I_1)$ given by Proposition 2.2.2 with frequency $(\omega_1/\sqrt{\varepsilon}, 0)$. It is however not clear that it exists and is unique in a $O(\sqrt{\varepsilon})$-neighbourhood of the origin in I_1. This again has to do with straightening and is an application of §1.9; we postpone the discussion to §2.5.1 and here simply collect the necessary hypotheses, assuming they are met.

We have to focus on relatively Diophantine frequencies because they survive by KAM theory, applied here to the torus with frequency $\omega = (\omega_1, 0)$ we started from in §2.1.4. But the same kind of remarks applies as for the nondegeneracy condition cast on the transverse Poincaré-Melnikov matrix. *If* one is able to prove the persistence of other tori or even other invariant objects by other means, *then* the discussion below also applies to these. For instance one can restrict the kind of perturbations one is working with by *imposing* that the whole chain of tori be preserved; this is the case in Arnold's original paper [A1] and will also be assumed in most of Chapter 3 for simplicity. It is important at this point to understand that the Diophantine condition (5) plays a dual role: on the one hand it enables one to prove the persistence of the torus using standard KAM-type techniques; on the other hand it provides a suitable (α, K)-nonresonant domain (see Definition 2.1.2 and (6)) over which one can normalize to an exponentially high order using resonant normal forms *and* the analyticity of the system (which could in principle be dispensed with as far as KAM techniques are concerned). We add that in some sense it is the *second* role of the Diophantine condition which is the more essential and deeply rooted in the nature of the problem.

We will work in a sector near the origin in the (ε, μ)-plane defined by ε small enough and $\mu \leq \mu(\varepsilon)$. We assume both parameters are positive for notational simplicity and we are of course especially interested in the defining value $\mu(\varepsilon) = \varepsilon^{-1}\|R_\varepsilon\|$, which is exponentially small with respect to ε. But it is interesting and clearer to keep the flexibility of leaving the dependence $\mu(\varepsilon)$ free for the moment, assuming only that $\mu(\varepsilon) = O(\varepsilon^2)$ for consistency. Yet of course exponential smallness properties of the splitting matrix will hang on the fact that we will ultimately apply the results with exponentially small values of $\mu(\varepsilon)$.

We now gather basic assumptions which are sufficient (but far from necessary) to ensure exponential smallness properties for the splitting matrix. These assumptions will turn out to hold true in our case, but could also lead to results in a different setting, for instance if the integrable part h of the Hamiltonian H is not fully nonlinear but the perturbation f is restricted to a suitable class of functions.

We keep the notation of §§2.4.3–2.4.4. The following assertions are then stated for ε small enough and $\mu \leq \mu(\varepsilon) = O(\varepsilon^2)$.

ASSUMPTIONS.
i) The invariant torus $\mathcal{T}_\varepsilon(0)$ for H_r, with frequency $(\omega_1/\sqrt{\varepsilon}, 0)$, can be continued into a family $\mathcal{T}_{\varepsilon,\mu}$ of invariant tori for $H_{\varepsilon,\mu}$ with the same frequency. Moreover $\mathcal{T}_{\varepsilon,\mu}$ is μ-close to $\mathcal{T}_\varepsilon(0) = \mathcal{T}_{\varepsilon,0}$;

ii) the invariant manifolds $\mathcal{W}_\varepsilon^\pm$ can be continued as graphs of functions: over the domains $V^\pm$, $\mathcal{W}_{\varepsilon,\mu}^\pm = \mathcal{G}r(dS_{\varepsilon,\mu}^\pm)$; $S_{\varepsilon,\mu}^\pm$ is μ-close to $S_\varepsilon^\pm$ in the C^2 topology on $V^\pm$;

iii) the graphs $\mathcal{G}r(dS_{\varepsilon,\mu}^\pm)$ intersect near $\mathfrak{T}_\varepsilon(0)$ (such an intersection defines a homoclinic trajectory biasymptotic to $\mathcal{T}_{\varepsilon,\mu}$ for $H_{\varepsilon,\mu}$).

We make some remarks about these assumptions. As mentioned already the existence of $\mathcal{T}_\varepsilon$ is best seen by straightening the unperturbed ($\mu = 0$) flow as in §1.9 (see §2.5.1). We then add the μ perturbation in order to check the validity of i) under the assumptions of this chapter; this is a matter of classical hyperbolic KAM theory (§2.5.2). The validity of Assumption ii) then follows more or less "in the wash" (§2.5.3). Here, as explained above already, we are of course not considering the invariant manifolds $\mathcal{W}_{\varepsilon,\mu}^\pm$ globally but only in the vicinity of the homoclinic trajectory. Paragraph 2.5.4 is devoted to checking the validity of Assumption iii).

2.4.7. We are finally in a position to state results about exponential smallness of the splitting matrix! In this subsection we keep the rescaled normalized variables, *i.e.* those which one gets starting from (1) and performing the near-identity normalizing transformation $\mathcal{C}_\varepsilon$ followed by (8) and (12). Retrieving the information in the original variables will then be easy, using Lemma 2.3.2; this we do in §2.4.8 below and we will also give a more geometric and coordinate-free statement in §2.4.9.

Again we need to pick a homoclinic point in order to be able to write down a splitting matrix. And to this end we first need a section which we obtain by perturbing the section $\Sigma_\varepsilon(0)$ of §2.4.4. Indeed using the same hypersurface $\tilde{\Sigma} \subset T^*\mathbb{T}^\ell$ which was considered there we define:

$$\Sigma_{\varepsilon,\mu} = \tilde{\Sigma} \cap \{H = H(\mathcal{T}_{\varepsilon,\mu})\}.$$

Consider a homoclinic trajectory $\gamma_{\varepsilon,\mu}$ as in Assumption 2.4.6 iii); we call $p_{\varepsilon,\mu}$ the intersection point: $p_{\varepsilon,\mu} = \gamma_{\varepsilon,\mu} \cap \Sigma_{\varepsilon,\mu}$.

We start with a statement which by now has become essentially self-evident but is nonetheless significant and the seed of further estimates on the exponential smallness of the splitting matrix.

PROPOSITION. *With the setting and notation as above suppose that Assumptions 2.4.6 hold true. Consider the splitting matrix $M_{\varepsilon,\mu}$ at the point $p_{\varepsilon,\mu}$ inside the section $\Sigma_{\varepsilon,\mu}$. Then:*

(19) $$M_{\varepsilon,\mu} = M_\varepsilon + O(\mu),$$

for any value of $\mu \leq \mu(\varepsilon)(= O(\varepsilon^2))$, where the matrix M_ε has the form (18).

Let us collect part of the information this statement carries. We simply apply the lemmas on spectral stability of §2.3.1. Here we are dealing with a two-parameter situation but the adaptation is straightforward. In fact we have placed ourselves in the simplest case, by assuming that $M_0^\perp$ is nondegenerate. This implies that 0 has multiplicity n in the spectrum of M_ε and is isolated in the spectrum by a finite amount for ε small enough. The lemma on the continuity of the eigenvalues applies and for $\mu \leq \mu(\varepsilon)$ we get that $M_{\varepsilon,\mu}$ has exactly n eigenvalues (with multiplicities) of size $O(\mu)$ as μ and ε tend to 0. Let $\Pi_{\varepsilon,\mu}^{(0)}$ be the corresponding spectral projector for $M_{\varepsilon,\mu}$, $\Pi_\varepsilon^{(0)} = \Pi_{\varepsilon,0}^{(0)}$ and $\Pi^{(0)} = \Pi_0^{(0)}$. Clearly we have in fact $\Pi^{(0)} = \Pi_\varepsilon^{(0)}$ for any small ε and this is just the projector onto the kernel of $M_0^\perp$, which in turn consists of the first n basis vectors. So we get the estimate:

$$\|\Pi_{\varepsilon,\mu}^{(0)} - \Pi_\varepsilon^{(0)}\| = \|\Pi_{\varepsilon,\mu}^{(0)} - \Pi^{(0)}\| = O(\mu), \tag{20}$$

which gives precise information on the location of the n-dimensional eigenspace over which $M_{\varepsilon,\mu}$ has norm $O(\mu)$.

There only remains to apply the above to the case $\mu \leq \mu(\varepsilon) = \varepsilon^{-1}\|R_\varepsilon\| = O(\exp(-w\varepsilon^{-a}))$ for some $w > 0$, $a > 0$, to get results of exponential smallness. These will be spelled out in the next two subsections in the original variables and in a more geometric way.

As for now we still would like to mention that there is a hoard of variants and amplifications that can be derived. They have to do with the nondegeneracy condition on $M_0^\perp$ or rather with the fact that it is not really necessary. As already mentioned, it is indeed useful in order to show that Assumptions 2.4.6 hold; but assuming that they do for such or such reason, the stability lemmas are quite flexible and their range of application can easily be enlarged. We will be content with just sketching a useful amplification and leave it to the reader to adapt or enlarge it further if need be.

So let us drop the nondegeneracy assumption on $M_0^\perp$ for a moment; consider a spectral projector Π_ε for M_ε (and ε small enough). Assume that Π_ε corresponds to a subset Λ_ε of the spectrum of M_ε such that the distance of Λ_ε to the rest of the spectrum is large with respect to $\mu(\varepsilon)$, whereas the size of Λ_ε is much smaller. The case above corresponds to just picking $\Lambda_\varepsilon = \{0\}$. For illustration, let us simply consider the practically important case where the size (in the obvious sense) of Λ_ε is polynomial, say on the order of ε^p with some $p \geq 1$, whereas $\mu(\varepsilon)$ is exponentially small and the rest of the spectrum is "far away", that is lies at a distance which is large with respect to ε^p. Let $\lambda_\varepsilon \in \Lambda_\varepsilon$ be a continuous determination of some of the elements of Λ_ε. Problems of labeling, crossings etc. are immaterial here; one does not even need that λ_ε be in Λ_ε, only that it lies in its convex hull (which is simply a polynomially small interval). Then we can blow up the spectrum around Λ_ε by considering the matrix $\varepsilon^{-p}(M_{\varepsilon,\mu} - \lambda_\varepsilon)$. The origin becomes isolated in the spectrum and we can apply the spectral stability lemmas as above. This can be useful even in the initial case when $M_0^\perp$ is nondegenerate, if one is interested in other parts of the spectrum for measuring the polynomially small transverse splitting. The moral of the tale can be summarized by saying that one can in fact resolve clusters in the spectrum on any scale down to (and excluding) that of μ. For instance if μ is exponentially small with exponent at least a (as will be the case in our context), one can even separate clusters which are exponentially close, provided the exponents of the mutual distances are $< a$.

Lastly we remark again that we have been dealing with the splitting matrix at a homoclinic point, *i.e.* the difference of the second derivatives of the generating functions for the invariant manifolds, but we could also consider the distance between these manifolds, *i.e.* the first derivatives of the generating functions, or for that matter the generating functions themselves. This will be exemplified in Chapter 3 below. But it may be fit to recall that there is an additional difficulty when dealing with *hetero*clinic problems, namely that the two invariant objects (tori) have to lie in the same chart for the normalizing coordinates. Here for instance using this symplectic method we can in any event simultaneously only examine tori which are $O(\sqrt{\varepsilon})$ apart in the original action coordinates, a distance which is however quite

2.4.8. Here we move back to the original variables in (1), still assuming however that the resonance we consider is in standard form, which means that the lattice $\mathcal{M} \subset \mathbb{Z}^\ell$ (cf. §2.1.1) is generated by the last m vectors of the standard basis. We will state the purely geometric statement in the next subsection, getting rid of this last minor assumption.

Applying the inverse of the scaling (8) amounts to multiplying the splitting matrix $M_{\varepsilon,\mu}$ by $\sqrt{\varepsilon}$. Then we still have to take into account the effect of the near-identity normalizing transformation $\mathcal{C}_\varepsilon$ (or rather its inverse). To this end we use Proposition 1.5.4 in this simple perturbative setting, as already discussed in §2.3.4. We could of course again state a result as in Proposition 2.4.7, which holds for any $\mu \leq \mu(\varepsilon)$ and any function $\mu(\varepsilon)$ provided it is small with respect to ε (see above). We leave this to the reader and in order to make things more suggestive we confine ourselves to the case we are especially interested in, namely $\mu = \mu(\varepsilon) = \varepsilon^{-1}\|R_\varepsilon\| = O(\exp(-w\varepsilon^{-a}))$ where $w > 0$ and $a = 1/(2\tau_1)$ are as in Proposition 2.1.4.

We let $\mathbf{M}_\varepsilon$ denote the splitting matrix for $(\varepsilon, \mu(\varepsilon))$ inside the section $\Sigma_\varepsilon = \Sigma_{\varepsilon,\mu(\varepsilon)}$ and using the original variables of (1). In other words this is really the object we were interested in from the start, except for a linear symplectic change of coordinates which has reduced the resonance to the standard form. This obviously leads us to the

PROPOSITION. *With the setting and notation as above suppose that Assumptions 2.4.6 hold true with $\mu(\varepsilon) = \varepsilon^{-1}\|R_\varepsilon\|(= O(\exp(-w\varepsilon^a)))$. Then the splitting matrix $\mathbf{M}_\varepsilon$ satisfies:*

$$(21) \qquad \mathbf{M}_\varepsilon = \sqrt{\varepsilon}(1 + O(\varepsilon))\big(M_\varepsilon + O(\exp(-w\varepsilon^{-a}))\big)(1 + O(\varepsilon)),$$

with the matrix M_ε as in (18); the exponent a is given by $a^{-1} = 2\tau_1$ and the width w is strictly positive.

Thanks to Lemma 2.3.2, or rather the perturbative version discussed there, this estimate immediately implies exponential estimates for the spectral data of $\mathbf{M}_\varepsilon$. Given two varying subspaces V_ε and V_ε' in Euclidean space, we will say that they are ε-close if the norm of the difference of the two attending orthogonal projectors is $O(\varepsilon)$. This is of course equivalent to the Euclidean measure of their angle. Assuming that $M_0^\perp$ is nondegenerate, we get the

COROLLARY. *The splitting matrix $\mathbf{M}_\varepsilon$ has exactly n ($= \ell - m$) exponentially small eigenvalues. Their exponents are larger than a ($= 1/(2\tau_1)$) and the corresponding eigenspace is ε-close to the subspace spanned by the first n basis vectors.*

This implies of course that the determinant of the matrix is also exponentially small, with exponent at least a, in fact equal to the largest exponent of the n exponentially small eigenvalues. Note that because of the nondegeneracy assumptions on $M_0^\perp$ there are exactly n exponentially small eigenvalues and no more, as again the min-max principle (Lemma 2.3.2) applied to the other nonzero eigenvalues of M_ε shows. The remarks about relaxing the nondegeneracy conditions still apply and in fact, using Lemma 2.3.2 to take care of the perturbation "of multiplicative type" in (21), one can study the spectrum of $\mathbf{M}_\varepsilon$ just as we outlined above for that of $M_{\varepsilon,\mu}$. We leave the details to the interested reader.

2.4.9. It may be useful to rephrase part of the above in a more geometric, coordinate-free way and this is what we briefly do in this last subsection. We certainly do not collect all the available information but rather try to express part of it in a more suggestive manner. In doing so we allow ourselves some slightly informal expressions, which however should be easy to clarify after the developments above.

Return to the initial setting in §2.1. So let again $\mathcal{M} \subset \mathbb{Z}^\ell$ be a lattice of rank m, with associated resonance surface $S_\mathcal{M}$ in action I-space. We assume that h is nondegenerate, so that the m-fold resonance surface $S_\mathcal{M}$ is a regular submanifold of dimension $n = \ell - m$. Let $I_0 \in S_\mathcal{M}$ and assume that the corresponding frequency $\omega = \omega(I_0)(= \nabla h(I_0))$ is relatively Diophantine (with respect to $\mathcal{M}$) with exponent $\tau_1 \geq n$. This means that $|k|^{\tau_1 - 1}|\omega \cdot k|$ is bounded away from 0 as k runs through $\mathbb{Z}^\ell \setminus \mathcal{M}$ (compare (5)). Assume further that the reduced first-order averaged system has a nondegenerate fixed hyperbolic point x_0 with sign-definite Hessian matrix (see under (14) and §2.2.2) and that there is a trajectory γ biasymptotic to x_0 (this is actually always the case; cf. §2.5.4 below). Finally assume that one can perturb this situation in the sense that Assumptions 2.4.6 hold for ε small enough and $\mu = \mu(\varepsilon) = \varepsilon^{-1}\|R_\varepsilon\|$. So we consider the perturbed trajectory from γ and a homoclinic point $\mathbf{p}_\varepsilon$ on it ($\mathbf{p}_\varepsilon = p_{\varepsilon,\mu(\varepsilon)}$ in the notation above), a section inside the energy surface and the corresponding splitting matrix $\mathbf{M}_\varepsilon$ at $\mathbf{p}_\varepsilon$, which is square symmetric of size $(\ell - 1)$. Then one has:

THEOREM. *With the assumptions as above the splitting matrix $\mathbf{M}_\varepsilon$ has exactly n exponentially small eigenvalues (counting with multiplicities). Their exponents are at least $a = 1/(2\tau_1)$ and the corresponding n-dimensional eigenspace is ε-close to the tangent space of $S_\mathcal{M}$ at the point I_0.*

PROOF. There remains only to consider the linear symplectic transformation which brings $\mathcal{M}$ into standard form. An application of Lemma 2.3.2 produces the statement above. This linear transformation has no effect on the exponent but it does change the width, an interesting and important point when looking at the global picture, *i.e.* the whole "web" of resonance surfaces or the counterpart in simultaneous approximation. The attending constants $c(\mathcal{M})$ are discussed in [L2] (Chapter III). □

Again, substantially more information is in fact available, as partly spelled out above, but we hope that this result illustrates the motto italicized in the introduction to this chapter. We remark that the sign "$\perp$" which we use (in $M_0^\perp$ etc.) is not an usurpation because the matrices are symmetric, so that the eigenspaces are orthogonal for the ordinary Euclidean structure. In that motto, the words "inside the energy surface" were understood, and should be restored. This goes along with taking a symplectic section which avoids considering the zero eigenvalue of the full splitting matrix arising from variation along the homoclinic trajectory. One may also simply not perform any section and study the full splitting matrix. However sections are certainly necessary if only the *determinant* of the splitting matrix is being studied, as was the case (and indeed only in very special cases) prior to the present paper.

2.5. Persistence of tori, invariant manifolds and homoclinic trajectories

This paragraph is concerned with discussing the validity of Assumptions 2.4.6 in the setting of the present chapter, thus showing the unconditional validity in this setting of the statements in §§2.4.7–2.4.9. Given all the preparatory work this will in fact be comparatively easy and we will be brief. We follow the plan outlined after the statement of Assumptions 2.4.6 and record the conclusions in §2.5.5.

We discuss only the fully nonlinear case, that is we assume that the Hessian matrix of the integrable part in (1) is nondegenerate. It is actually enough to make this assumption at the origin $I = 0$, which is the same as assuming that the frequency map $I \mapsto \omega(I)$ is a local diffeomorphism near the origin. This could be weakened and it is clear that a large combinatorics of variants exists but we will not go into this here. In fact we also assume that the partial Hessian matrix B (see below (14)) transverse to resonance is sign definite, so that in particular Proposition 2.2.2 applies.

2.5.1. Straightening the resonant flow.
Here we use §1.9.2 in order to straighten the invariant manifolds when passing from H_r^0 to H_r. Concretely this eliminates the "mixed" terms in the normal form (see §1.11.2 and §2.2) and facilitates the application of KAM theory, namely of the results in [Gr] and [T1].

Let $\mathbf{N}$ be the normally hyperbolic invariant manifold for H_r which is the union:

$$\mathbf{N} = \bigcup_{|I_1|<\rho/2} \mathbb{T}^n \times \{(I_1, x_\varepsilon(I_1))\},$$

where $x_\varepsilon(I_1)$ stands for the hyperbolic fixed point of the system H_r when fixing the n first action variables I_1. Here of course we always assume that ε is small enough and we change ρ into $\rho/2$ because the fixed points move slightly with ε and we need that they stay in the domain where the normal form is valid. Below in this pararaph we will omit the mention of similar details.

The manifold $\mathbf{N}$ corresponds to $\mathbf{N}_0$ in §1.9.2; we omit the subscript 0 referring to $\mu = 0$ because we won't need to consider the perturbed version of $\mathbf{N}$ for nonzero μ, although of course it does exist, just as in §1.9.2. We denote by $\mathbf{W}^\pm$ the invariant manifolds of $\mathbf{N}$, so that $\mathbf{N} = \mathbf{W}^+ \cap \mathbf{W}^-$. Following §1.9.2 we perform a straightening transformation $\mathcal{S}_\varepsilon$ which is ε-close to identity. We denote the new variables by $(J, \psi, s, u) \in \mathbb{R}^n \times \mathbb{T}^n \times \mathbb{R}^m \times \mathbb{R}^m$:

$$(J, \psi, s, u) = \mathcal{S}_\varepsilon(I, \phi).$$

In the new system of coordinates both $\mathbf{N}$ and the manifolds $\mathbf{W}^\pm$ have been straightened: the equation for $\mathbf{N}$ is $s = u = 0$, those for $\mathbf{W}^+$ and $\mathbf{W}^-$ are $u = 0$ and $s = 0$ respectively. We may assume moreover (cf. §1.9.2) that the torus $\mathcal{T}_\varepsilon(0)$ is given by $J = 0, s = u = 0$, with invariant manifolds $\mathcal{W}_\varepsilon^\pm$ given by $J = 0, u = 0$ and $J = 0, s = 0$ respectively.

Now in the new system of coordinates, the resonant part H_r can be written (cf. §1.11.2, (N)):

$$H_r \circ \mathcal{S}_\varepsilon = K_r(J, s, u, \varepsilon) = \frac{\omega_1}{\sqrt{\varepsilon}} J + \Delta(\varepsilon) J^2 + s \cdot \Lambda(\varepsilon) u + O_3(J, s, u; \psi).$$

Here $\Delta(\varepsilon)$ is an n-symmetric nondegenerate matrix and the Lyapunov matrix $\Lambda(\varepsilon)$ is an m-symmetric positive matrix. Because of the scalings (8), this matrix is of order 1. Note that if we denote by L_r the difference $L_r = K_r - \frac{\omega_1}{\sqrt{\varepsilon}}J$, with an obvious extension for $\varepsilon = 0$, then $L_r(\varepsilon) = L_r(0) + O(\varepsilon)$.

2.5.2. Finding the tori: the KAM step. Here we use classical KAM techniques in order to prove that some tori persist when the μ-perturbation is turned on, thus checking the validity of Assumption 2.4.6 i) in our setting. Note that this is the only place in this chapter where KAM techniques are used or at least invoked. Of course one can and often does consider perturbations which simply leave the unperturbed tori invariant, as will indeed be the case in Chapter 3 (see however §3.6.2). If so, just move to the next item ...

The mixed terms in the resonant part H_r have been eliminated by the straightening transformation $\mathcal{S}_\varepsilon$ performed in the last subsection and we have actually reached a normal form of the type considered in §1.11.2: one can compare with (N) there, with A and Λ independent of the nonresonant angles ϕ. We note that an alternative possibility would have been to just keep those mixed terms in H_r, skipping the straightening step of §2.5.1 above. In principle these terms could be dealt with directly as in [R1] and [R3]. Weaker nondegeneracy conditions to be imposed on the integrable part h are also provided by this last paper.

In the (J, ψ, s, u)-coordinates, the full Hamiltonian can be written:

$$H \circ \mathcal{S}_\varepsilon = K(J, \psi, s, u, \varepsilon) = K_r(J, s, u, \varepsilon) + \mu R(J, \psi, s, u, \varepsilon).$$

We may now apply the results of [Gr], or rather those of [T1] which is an adaptation of [Gr] to our case, in order to continue the tori $\mathcal{T}_\varepsilon(0)$ into a two-parameter family $\mathcal{T}_{\varepsilon,\mu}$ for ε small enough and $\mu \leq \mu(\varepsilon)$. The only differences of the present setting with that of [T1] are as follows. First in [T1] the scalings (8) are not performed so that the frequencies are on the order 1 and the Lyapunov exponents on the order of $\sqrt{\varepsilon}$; this is of course completely inessential for the present purpose. Second in [T1] there is only one small parameter ε and the integrable part is ε-independent. But because of what was noted above (namely $L_r(\varepsilon) = L_r(0) + O(\varepsilon)$; see §2.5.1), it is plain that the proof simply goes through. Moreover it is actually valid for μ *polynomially* small with respect to ε ($\mu = O(\varepsilon^2)$ would do), so *a fortiori* if this last parameter is exponentially small. Lastly the invariant torus $\mathcal{T}_{\varepsilon,\mu}$ is μ-close to $\mathcal{T}_\varepsilon(0)$, since the effective perturbation $H - H_r$ is of size μ.

2.5.3. Invariant manifolds. As an appendix to the last subsection we briefly mention, using [Gr] and [T1] again, the fate of the (local) invariant manifolds attached to the invariant tori which were spotted using KAM theory. This enables us to check the validity of Assumption 2.4.6 ii). In fact there is now very little to be said. Working as in [T1], which follows [Gr] on that matter, one gets that the invariant torus $\mathcal{T}_{\varepsilon,\mu}$ has invariant manifolds $\mathcal{W}^\pm_{\varepsilon,\mu}$, which are actually straightened together with the (perturbed) flow on them.

This means that there exists a *fixed* relatively compact neigborhood B of $\mathcal{T}_\varepsilon(0)$ with the following properties: on B there is a a system of coordinates (J', ϕ', s', u'), with $(J', \phi', s', u') = \mathcal{R}_{\varepsilon,\mu}(J, \psi, s, u)$, where the canonical transformation $\mathcal{R}_{\varepsilon,\mu}$ is defined on B and μ-close to the identity and such that in these coordinates the torus $\mathcal{T}_{\varepsilon,\mu}$ has equations $J' = 0$, $s' = u' = 0$ and the local stable and unstable manifolds $\mathcal{W}^\pm_{\varepsilon,\mu} \cap B$ have equations $J' = 0$, $u' = 0$ and $J' = 0$, $s' = 0$ respectively. Moreover there is a function $\Lambda(s', u', \varepsilon, \mu)$, defined on B with values in the symmetric positive definite m-matrices and such that the flow on the local stable manifold

$\mathcal{W}^+_{\varepsilon,\mu} \cap B$ is given by:

$$\frac{dJ'}{dt} = 0, \qquad \frac{d\psi'}{dt} = \frac{\omega_1}{\sqrt{\varepsilon}}, \qquad \frac{ds'}{dt} = -\Lambda(s', u', \varepsilon, \mu)s', \qquad \frac{du'}{dt} = 0;$$

on the local unstable manifold $\mathcal{W}^-_{\varepsilon,\mu} \cap B$ it is given analogously by the equations:

$$\frac{dJ'}{dt} = 0, \qquad \frac{d\psi'}{dt} = \frac{\omega_1}{\sqrt{\varepsilon}}, \qquad \frac{ds'}{dt} = 0, \qquad \frac{du'}{dt} = \Lambda(s', u', \varepsilon, \mu)u'.$$

This being said, in order to check the validity in our setting of Assumption 2.4.6 ii), it is enough to note that the domains $V^\pm$ there are relatively compact, and can thus be covered in projection by flowing the local invariant manifolds $\mathcal{W}^\pm_{\varepsilon,\mu} \cap B$ for a finite fixed time. This last assertion comes from the expression of the perturbed flow on them and the fact that the matrix $\Lambda(s', u', \varepsilon, \mu)$ is of order 1; indeed it is actually μ-close to the matrix $\Lambda(\varepsilon)$ appearing in the expression of the resonant part K_r in §2.5.1 above.

2.5.4. Existence of homoclinic trajectories. We first consider again the generalized pendulum P. Using [St] and [CS] one can show that there are actually at least m homoclinic trajectories attached to the hyperbolic point x_0. However we have to assume in addition that at least one is nondegenerate, *i.e.* the corresponding transverse Poincaré-Melnikov matrix is nondegenerate, which is a generic condition. We pick once and for all such a nondegenerate homoclinic trajectory γ.

We keep the notation of §2.4.4 and the problem is to show that the manifolds $\mathcal{W}^\pm_{\varepsilon,\mu}$ viewed as the graphs of the differentials $dS^\pm_{\varepsilon,\mu}$ over $V^\pm$ do intersect. This is an application of §§1.10.1–1.10.2. Indeed using the notation of the present paragraph, it is shown there that upon adding the μ-perturbation there emerge from the torus $\mathcal{T}_\varepsilon(0)$ at least $(n+1)$ homoclinic points (in the section $\Sigma_{\varepsilon,\mu}$). In the generic case where the invariant manifolds $\mathcal{W}^\pm_{\varepsilon,\mu}$ intersect transversely (still in $\Sigma_{\varepsilon,\mu}$), one can label and follow such a homoclinic point as $p_{\varepsilon,\mu}$ and then the splitting matrix considered, say, in Theorem 2.4.9 pertain to such points, with $\mu = \mu(\varepsilon)$. But strictly speaking this last genericity assumption can be dispensed with and the conclusions of the statements in §2.4 are valid for *any* homoclinic point with ε small enough and $\mu = \mu(\varepsilon)$.

2.5.5. Here we simply record as a theorem the conclusions we have reached. The assumptions on the system we are looking at are as above. In particular the origin $I = 0$ is assumed as usual to correspond to a relatively Diophantine frequency with respect to a lattice $\mathcal{M}$ of rank m. Now apart from the assumptions stated *e.g.* before Theorem 2.4.9, we add nondegeneracy conditions as used above, which we repeat as follows:

i) the Hessian matrix $\partial^2 h/\partial I^2(0)$ of the integrable part h at the origin is assumed to be nondegenerate;
ii) the partial Hessian matrix with respect to the slow variables (*i.e.* $B = \partial^2 h/\partial I_2^2(0)$ after normalizing the resonant lattice $\mathcal{M}$) is strictly definite (positive or negative);
iii) the function $g = Z(0, \phi_2, 0)$ (*cf.* (4)) on $\mathbb{T}^m$ obtained by averaging the perturbation has a nondegenerate maximum (resp. minimum) if the matrix considered in ii) is positive (resp. negative), and we denote by x_0 the corresponding hyperbolic fixed point;

iv) consider one of the homoclinic trajectories $\gamma \in T^*\mathbb{T}^m$ attached to x_0; we assume that its stable and unstable manifolds intersect transversely in their energy level.

Note that the homoclinic trajectories considered in condition iv) do exist as mentioned in §2.5.4. Then we have proved:

THEOREM. *Under the above conditions i) to iv), Assumptions 2.4.6 hold true, so that the conclusions of Theorem 2.4.9 are valid in this case.*

Note that of course the statements in §§2.4.7–2.4.8 are valid just as well; all these statements are not empty in the sense that there do exist homoclinic points, as shown above.

2.6. Splitting and stability

There are actually—at least—*four* types of exponential quantities this paper is directly or indirectly concerned with and they give rise to four types of exponents, namely the linear splitting, splitting, stability and instability exponents. On top of this each of these quantities has a corresponding local version. In the best of all worlds all four quantities have the same order of magnitude and in particular all four types of exponents should coincide. But there are many qualifications and caveats around before the above can even be phrased in a convincing way. In this paragraph we will essentially discuss the connection between stability and splitting, first recalling some important facts about stability over exponentially long times for near-integrable systems. We will hardly touch the connection between the size of the splitting, long time stability and the speed at which instability (Arnold diffusion) can develop. The reader can find indications on this in [L2] and [L4] and it is to be hoped that future works will make this difficult point clearer. In fact more information, especially of geometric nature can already be gathered from [B], [Be1,2,4], [Cr1,2] and [Mar1,2]. We also mention that the connection between splitting and linear splitting (Poincaré-Melnikov approximation) will be discussed in Chapter 3 (see in particular §3.5). Clearly the historical tour we propose below is much too detailed for our immediate purposes and the hurrying reader is welcome to skip it and jump to §2.6.6. But it may be useful to recall some of the motivations of the theory, how it developed and possible open paths. And in fact on the way we present some new problems and sketch some new results, even if they are not stated and proved as such.

2.6.1. We start with an historical and cursory review on long time stability in near-integrable Hamiltonian systems, revisiting some landmarks and insisting on some physical or simply heuristic motivations and principles. We look back at the general near-integrable Hamiltonian (1) and its study, which Poincaré, in an oft-quoted sentence from [P], designated as the "general problem of dynamics", partly perhaps, one is tempted to add, because others just seem out of reach. What we would like to call *classical perturbation theory* is largely concerned with comparing the trajectory of the perturbed system with that of the averaged system. Now for Hamiltonian systems the averaged motion of the action variables vanishes to all orders. In fact Hamilton's equations for these variables read: $dI/dt = -\partial H/\partial \phi$, which has zero average on the torus for any H. This reduces the problem to that of the *stability* of the action variables. So starting from (1) with initial point $(I(0), \phi(0))$, we want to show that $\|I(t) - I(0)\|$ remains small for a long time $T(\varepsilon)$, and *any* initial value $I(0)$.

It is quite remarkable that the problem above was actually posed in this general way, and essentially solved, only quite recently. We will certainly not give an orderly review (see [N], [L2], [L3]) but just recall that it is especially meaningful to consider *analytic* and *steep* perturbations, which lead to exponentially large stability times $T(\varepsilon)$. Note that we use the terms *exponents*, *width* etc. for an exponentially large quantity by referring to the inverse, exponentially small quantity (see the introduction to this chapter); so we have $T(\varepsilon) \succeq \exp(w\varepsilon^{-a})$ for some $w > 0$ and $a > 0$. This is the central result in [N], which was actually obtained (and announced in a short note) around 1970.

Historically speaking, it is truly worthy of notice that prior to that date, there did not exist results of this type even over *polynomially* long times. As is well-known Kolmogorov proved the existence of invariant tori in 1954 and *geometric perturbation theory*, *i.e.* the search for invariant objects in phase space, was then developed much more actively than the equally physically relevant classical perturbation theory. It should however also be pointed out that the methods are not so terribly different, as perhaps is illustrated in the present paper. Meanwhile, that is between 1954 and 1970, V.Arnold produced his example of global instability ([A1]), about which he writes (in [A2]): "My main contribution [to dynamics] was the discovery (in 1964) of the universal mechanism of instability in the systems with many degrees of freedom, close to integrable, — later called 'Arnold diffusion' by the physicists". We propose (as in [L4]) to call this mechanism and its variants "Arnold mechanism", but the phrase "Arnold diffusion" has certainly produced in its vagueness many ambiguous or nonsensical statements. In any case Arnold mechanism is still to-date in essence the only one we know of for producing geometric instability in close-to-integrable systems. We mention that after reading [K] and [A1] (8 pages in all), perhaps with the help of [BGGS] and [L4] for "details", the newcomer will have learned a very sizable portion of the whole subject.

Returning to Nekhoroshev's 1977 paper (again the results were obtained much earlier), it introduced (at least) two very important ideas, namely resonant normal forms to an exponentially high order on the side of analysis, and the puzzle or patchwork which is used in order to control the geometry of the resonances. He proved a global (in action space) stability result for steep analytic near-integrable systems. Of course the main thrust of the paper is that the exponents exist at all in the analytic steep case, *i.e.* the discovery of stability over exponentially long times. Coming to the *values* of the exponents, if the unperturbed Hamiltonian is assumed to be convex (h convex in (1)), Nekhoroshev obtained a value of the exponent a on the order of $1/\ell^2$ and there is no conjecture in his paper about a possible optimal value. We insist on the fact that at the time of this writing, *there is still no conjecture (much less proof) about the optimal value of the stability exponent(s) in the steep non quasi-convex case*. In other words, although steepness exponents have a nice geometric interpretation in the analytic setting (due to Y.Illiashenko, A.Neishtadt and N.Nekhoroshev), we do not as yet know the exact connection between the stability and steepness exponent(s). To be completely explicit, we stress the obvious fact that by its very definition, the value of a gives a rough *lower* estimate of the time or say timescale over which instability can occur at all. Here "instability" is intended in a fairly weak sense, as a drift of order 1 in the action variables, uniformly as the perturbation parameter goes to 0; this is actually the definition used in [A1]. The point is that optimizing the value of a is not an abstruse matter for picky specialists but rather *it defines the limits (in time) of*

Hamiltonian perturbation theory. Beyond that timescale, the theory simply breaks and one enters a new regime where global instability has to be taken into account in a crucial way.

2.6.2. As is often the case, important information was rather slow to circulate and it took many years before Nekhoroshev's result was noted and disseminated in what was still the western world. Saying a few words about this story may be more than just entertaining. First one may notice that there are many exponential functions which appear in physics (and hopefully in Nature). It can be argued that the exponential function is probably the simplest transcendental function so that there are several purely formal reasons why it may show up. This makes it indeed difficult to decide whether certain similarities are deeply rooted in the nature of things or simply formal. To name but a few important instances, exponentials govern the decay of the Fourier coefficients of analytic functions, they appear in Planck's law for the black body radiation and in Boltzmann's law for energy repartition, but also in the law of large numbers and in Maxwell's velocity distribution in a diluted gas. Some of these phenomena are known to be related, although of course it took genius to see it: this is now rather obvious for the last two and one may recall that Planck's law follows if one uses Boltzmann's statistics by considering the black body radiation as a gas of photons (a "computation" which first appears in de Broglie's thesis). It so happens that Boltzmann had put forth an hypothesis (in 1895!) to explain the failure of energy equipartition (freezing of certain high frequency degrees of freedom) which was observed already in classical statistical mechanics. But the "ultraviolet catastrophe" was resolved (or avoided) shortly after by the mysterious introduction of the quanta by Planck and Boltzmann's hypothesis of a classical dynamical cut-off sank into oblivion. Long after, from 1952 to the death of E.Fermi in 1954 (the very same year Kolmogorov published his theorem on invariant tori), Fermi, Pasta and Ulam, working on a computer which had only slightly evolved since it had been used for the planning of D-day, found that energy equipartion fails in the nonlinear string which now bears their names. They were actually trying to check a famous prediction by Einstein on heat conduction (which of course is largely true all the same!) and failure to do so was certainly fruitful, as it marked the birth of what was to be called the "soliton", another story which is again tightly connected both to what appears below and to Kolmogorov's theorem. In fact it seems that a complete explanation of the Fermi-Pasta-Ulam numerical experiment is actually quite subtle, as it mixes several aspects: one finds the Korteweg-de Vries equation by going to a particular continuum limit, which is a form of singular perturbation; this equation is completely integrable, meaning in particular that it has many quasi-periodic solutions (multisolitons, assuming periodic boundary conditions), but the perturbation theory of the KdV equation is quite subtle and as yet incomplete. If one starts again from the original chain, without going to the continuum limit, one can develop a kind of KAM-type theory, but also study the long time stability; yet one should beware of the fact that the Fermi-Pasta-Ulam chain is a perturbation of *linear* oscillators, not rotators, *i.e.* the integrable part is isochronous which is quite special, both in the perspective of finding invariant tori and from the viewpoint of classical perturbation theory.

In any case, advancing into the early eighties, we now meet with a group of physicists in Milano around L.Galgani, who having read Nekhoroshev's paper which at that time was virtually unknown (although the result was more than ten year old already), thought that it might be used to revive Boltzmann's proposal, which had been taken quite seriously for some time but had not resisted the advent of quantum techniques. This was the starting point of yet another beautiful story, which again as hopefully transpires below is very much connected with our main concern and with the present paper: we refer to [BGG2] for a concise exposition of the physical motivations; other references can be found in many places, included an incomplete list in [L3]. Again one main point was to find a purely classical reason for the failure of energy equipartition which is observed over human times say. In principle this could open the way to a purely classical derivation of Planck's law (this may be hard to believe but ...) and also to a novel approach to some problems of statistical mechanics.

Coming slowly back to mathematics, there was from the start a clear concern in Milano about the fact that the stability exponent $a = a(\ell)$ for convex systems (the proof in the general steep case has not been revisited in print since [N]) was decreasing rapidly with the dimension ℓ indeed like ℓ^{-2}, as found in [N] and confirmed in [BGG1] (and other papers as well). And in any case for the application to statistical physics, the naive value was simply $\ell = \infty$ ... Lots of work was devoted at that time to the problem, including numerical computations on toy models for statistical mechanics, namely nonlinear one-dimensional chains of rotators and oscillators in the Fermi-Pasta-Ulam style. Again references are easily available and this is still an active subject of research, but we want to mention the fact that an important physical property slowly emerged, to wit that *resonance can stabilize the motion*, in the presence of nonlinearity (so this does not contradict parametric resonance) but in the absence of friction. Very roughly, one observed numerically that in reversible nonlinear systems there are cases where a resonant state survives for an *a priori* surprisingly long time, whereas nonresonant states occur as transients. One should bear in mind that, in the context of these models of statistical mechanics, resonance is the same as (nonlinear) localization, in other words action space and physical space somehow coincide and being located on a resonant surface of finite *dimension* (but infinite multiplicity) simply means juggling a finite number of beads in the chain. This can now be taken as an indication that the global stability exponent a does not tell the whole story, although of course it was not formulated in this way at that time. But the Milano story certainly ties up with the one in Los Alamos we have briefly alluded to.

2.6.3. Let us make a last instructive detour into the rather grim Soviet Union of the seventies, returning to the consideration of finite-dimensional near-integrable systems. To reduce a long, manifold and interesting story to just a few words, let us simply mention that in the fresh crisp air of Siberia and under the original and colourful impulse of the physicist V.Budker, a knowhow on "stability" was developed, with the stability of strongly focussing accelerators as one of the primary goals (prudently sticking to civilian applications). This distant colony of the Moscow (or say European Russian) school of nonlinear physics (one of the landmarks being the famous 1958 book by N.Bogoliubov and Y.Mitropolski on "Asymptotic methods in the theory of nonlinear oscillations"; see the bibliographical note in [LM] for a short link with today's views) flourished and found its own style. In particular B.Chirikov and coworkers initiated important physical and numerical studies on finite-dimensional near-integrable systems. This resulted in a long series of preprints from the Institute of Nuclear Physics in Novossibirsk, many not so easily available and spanning more than two decades. An important summary of the information is available in [C], in which one finds a conjecture about a possible *lower* bound for the speed of diffusion. It is actually based on an—informal and conjectural—connection between the speed of diffusion and the size of the splitting, and was formalized (not proved of course!) in [L2] (§V.2), at least in the fully nonlinear case (see also [L4]).

We stress that this conjecture is about a *lower* bound for the speed of diffusion. This *ipso facto* implies an *upper* bound for the stability time, so it prescribes a *maximal* value for the stability exponent. Chirikov's reasonings have a strong physical flavour and in particular often use concepts pertaining to stochastic systems (starting with the word "diffusion") in a purely deterministic context; his papers are certainly not easy to read for "mathematicians", as for instance no small divisors or Diophantine conditions ever occur, so that they have to be restored at appropriate places. In particular Paragraph 7 of [C] which deals with multidimensional systems does not seem to have gained wide readership, yet remains quite interesting. In particular it contains the conjecture not only for fully nonlinear systems, but also in some mixed cases. We are now going to "translate" and discuss the counterpart in terms of *stability* results, which has not been done to-date; it turns out that this leads to interesting considerations, including open problems.

In numerical investigations, it is relatively easy to add dimensions by just adding (quasi)frequencies, and this is partly why one is lead to consider mixed problems with Hamiltonians of type:

$$H(I, \phi) = \omega_2 \cdot I_2 + h(I_1) + \varepsilon f(I_1, \phi).$$

Here we (almost) follow the by now usual notation in the present paper: $(I, \phi) \in \mathbb{R}^\ell \times \mathbb{T}^\ell$, the variables with index 1 (resp. 2) have dimension n (resp. m) etc. The data are assumed to be analytic over some domain in I_1 (and globally in ϕ) and ω_2 is supposed to be Diophantine with exponent $\tau \geq m$ (i.e. $|k|^{1-\tau}|\omega_2 \cdot k|$ is bounded away from 0 as k runs through $\mathbb{Z}^m \setminus \{0\}$). Lastly we assume that h is a *convex* function. The Hamiltonian H of course simply describes a quasiperiodic perturbation of an n-dimensional fully nonlinear integrable, indeed convex system h, together with a Diophantine condition on the quasifrequency. Note that Hamiltonian (∗) contains a particular case of this, which one obtains by specializing the torsion to $\alpha = (\alpha_1, 0)$ (with all the components of α_1 strictly positive) and by localizing near a simple resonance, or just picking F independent of (p, q), which uncouples the pendulum part. Also one may take a polynomial value for μ, say simply $\mu = \varepsilon^2$. The KAM theory for these systems is taken care of in the appendix, in the form of Theorem A.1.3 and Proposition A.1.4. We may now "translate" Chirikov's prediction in this case into the

CONJECTURE. *The stability exponent for Hamiltonian H above is at least equal to $1/(2f)$, with $f = \tau + n$.*

This number f (or, say, our interpretation of it), Chirikov calls the number of "linearly independent (incommensurable) frequencies". We refer to [CV2] in which Ω plays the role of our ω_2 and $Q - 1$ is our f; the authors do not bother to add arithmetic conditions but they are obviously necessary. We stress that this conjecture concerns the *stability* exponent. Needless to say we believe that this exponent is often *optimal* although this is certainly not always the case as the analysis in §2.6.4 below demonstrates; but showing the coincidence of the stability and the instability exponents is another delicate matter.

The first remark is that *to-date the conjecture above is open for $m > 1$*. We will return to this but first comment on the $m = 0$ case. Then we are back to the analytic perturbation of an integrable convex system and Chirikov predicted the value $a = 1/(2\ell)$ for the stability exponent ([C], end of §7.4), having in mind that this was also the value of the *instability* exponent. Actually he was starting rather from the point of view of splitting and instability (see [C] and [L2]) but this is not really the point here. For more than a decade he was worried about the discrepancy between his prediction and the $O(\ell^{-2})$ rigourous result obtained by Nekhoroshev and later confirmed in other papers. One can compare (7.39) in [C] with the discussion following (7.47). He wrote repeatedly about the problem; we refer to [CV1,2] which make interesting reading. Just be aware of the fact that Chirikov actually uses the large "adiabaticity parameter" λ, which satisfies $\lambda \sim \varepsilon^{-1/2}$. It seems that the difference in the respective languages prevented any kind of understanding on the part of mathematicians or even "mathematician-physicists" in spite of the fact that the prediction was obviously quite important or even fundamental, as again it is tantamount to a prediction on the absolute limit of applicability of *any* Hamiltonian perturbation theory whatsoever (for nonlinear systems; see below about the linear situation).

In any case that dilemna was resolved in [L2], where it is shown that Chirikov's prediction was indeed correct, as far as the *stability* exponent is concerned, and for perturbations of convex Hamiltonians (again there is no prediction available at present in the more general steep situation). So the rigorous upper bound which one can get from perturbation theory at last essentially matched the conjectured optimal value predicted by Chirikov, which should coincide with the *instability* exponent (see [L3] for more details). The proof in [L2] introduces and uses simultaneous approximation, bypassing most traditional ingredients of perturbation theory, among which small divisors and Fourier series, and in fact it is that method which made it clear why the value predicted by Chirikov is indeed quite natural, on top of providing an easy proof. Subsequently it was shown by J.Pöschel (in [Pö1]) that one can retrieve the result by refining the geometric part of the original scheme of Nekhoroshev although this makes it less transparent why the value thus obtained could be reached and is likely to be optimal.

2.6.4. Let us take a closer look at the conjecture above in the mixed cases and then turn to the consideration of the *local* versions of the exponents, which are strongly related both to what we have done in this chapter and to some of the stories we barely outlined. So the conjecture is proved in the $m = 0$ case but actually also in the case $m = 1$ because periodic perturbations of convex functions are quasi-convex (see [L2], p. 73; for the same reason it is open if $m = 1$ and h is quasi-convex but not convex). On the other hand, as already indicated, it is open for $m > 1$, in which case the integrable part is certainly not steep. We make two remarks. The first is actually (the sketch of) a result for a fairly large class of systems, namely the "mixed properly degenerate case"—admittedly an awful albeit accurate terminology. Indeed consider a Hamiltonian of the form:

$$H(I, \phi) = \omega_2 \cdot I_2 + h_1(I_1) + \varepsilon h_2(I_2) + O(\varepsilon^2),$$

where h_1 and h_2 are convex and this time ω_2 is *any* m-vector (no arithmetical condition). This is clearly a particular case of the above, but in fact it already covers a rather large class. The point is that one may in fact perform a finite number k of steps of normalization and the effective hypothesis is that there appears a convex nonlinear perturbation after this process. The above corresponds to $k = 1$ but hypotheses could be weakened. Then the techniques of [L2] (see p. 94 and [L5] for the case $n = 0$) would probably allow one to prove rather effortlessly (exercise) that H has stability exponent $a = 1/(2\ell)$ ($\ell = m + n$), i.e. just the same exponent as in the fully nonlinear convex case. These problems are also closely related to the question of the long time stability of elliptic fixed points to which the same technique was also successfully applied (in [Ni2] and [Pö2]). Returning to the Hamiltonian above, we note that the unperturbed part is not convex, not even the first-order part *i.e.* including the h_2 term (because there are two possible signs of ε), but the fact that there are two different scales effects a kind of decoupling. This is a case where the nonlinearity is somehow "hidden" not too far behind; if one assumes that ω_2 is τ-Diophantine the conjecture holds true since $\tau \geq m$ but it does *not* predict the optimal exponent if $\tau > m$. This illustrates how this exponent is best given by the fully nonlinear theory which is looming behind.

Next, and this is the second remark, we point out a difference between the linear (isochronous) and nonlinear (anisochronous) theories (*cf.* also [L2] §IV.2). Consider Hamiltonian (∗) with $m = d$, $n = 0$ and write $\omega_2 = \omega$. Were it not for the $\frac{1}{2}p^2$ term, which is the only nonlinearity left in the unperturbed system, one would be facing the perturbation of uncoupled oscillators described by the Hamiltonian $\omega \cdot I$. This linear case can be given a completely algebraic treatment, by studying the growth of the coefficients of the Birkhoff series directly from the combinatorics. This results in Gevrey estimates which were intensively studied in the eighties; part of this work is summarized in [F], to which we refer. The upshot is that one can prove stability estimates with exponent $1/\tau$ and the point is the factor 2 which enters when leaving the purely linear world, where no geometry at all is necessary, the frequency vector being constant. In other words adding nonlinearity in just *one* degree of freedom let the whole exponent drop rather dramatically by a factor 2. This is also visible in the splitting case: in [S] and [DGJS] the authors work with (referring to (∗)) $d = m = 2$, $\tau = m = 2$ and they find indeed the splitting exponent $1/(2\tau) = 1/4$ (taking good care of some trivial scalings to have the notation of these papers match ours). What the above remarks and results seem to tell us is that adding in nonlinearity can be stabilizing, sometimes more than strongly detuned linearity.

2.6.5. Let us turn to the *local* exponents, advancing towards a comparison with splitting exponents. Local exponents were introduced in [L2] (Corollary 2 p. 85; see also [L3] §3) and their primary virtue is to embody the "stabilization *via* resonance" which we have mentioned in connection with the Milano story (§2.6.2 above). They give a local version of stability which says roughly that inside a resonance strip (of width $O(\sqrt{\varepsilon})$) associated to a resonant surface of dimension n, the stability exponent is at least $1/(2n)$, and this value is conjectured to be optimal. The point is that the exponent depends on the dimension of the resonant surface, *not* on the global dimension of the ambient phase space. This was discovered again using simultaneous approximation and it comes for free in that way. It amounts to the mathematically trivial remark that if $\omega \in \mathbb{R}^n$, the simultaneous approximation theory of ω is *identical* to that of a vector $(\omega, 0)$ where 0 denotes the null vector of arbitrary dimension m (the multiplicity of the resonance). Together with the optimization of the global exponent, this has immediate and perhaps important consequences in celestial mechanics; we refer to [L2] (§IV.1) for a prospective discussion and to [Ni1] for a detailed application giving the best rigorous estimates to-date for the stability in the general planetary problems; these estimates could perhaps be pushed to realistic values in a computer-assisted way.

Coming back to statistical physics, a very important feature is that one can in fact take $m = \infty$, and thus also the global dimension $\ell = m + n = \infty$. In other words this applies to infinite-dimensional systems, in particular the nonlinear chains which we have encountered in Los Alamos and Milano. This application to nonlinear localization, which was one of the seeds in the introduction of simultaneous approximation, is outlined in [L2] (§IV.3) which however left many serious obstacles to overcome. A complete version (probably not the only possible one) appears in [BG]. Since then similar considerations were successfully applied in a PDE context (see [Ba] and [Pö3]).

Let us return to finite-dimensional systems. Actually stabilization *via* resonance appeared in Milano but also in Novossibirsk in a finite-dimensional context. Recall that Chirikov conjectured that (using our terminology) the stability exponent should be $1/(2f)$, with f the number of "linearly independent (incommensurable) frequencies". This is in fact a *local* prediction! Namely again near a resonant surface of dimension n and any multiplicity whatsoever, there are n such frequencies available. We leave it to the reader to put this in practice and state the local case of Conjecture 2.6.3. In fact by now it is essentially equivalent to prove the local or the global conjecture. How to pass from one to the other should be easy and this is why we simply stated the global case, insisting rather on the new feature represented by the mixed linear-nonlinear feature.

We stress at this point that *because of these local results* there remains, even at the conjectural level, a certain gap between Chirikov's prediction for the instability rate (in the fully nonlinear case say) and the results appearing in [L2,3] and [Pö1]. It comes from the fact that Arnold mechanism requires working in the vicinity of a simple resonance (Chirikov's "guiding resonance"); variants are possible but they also involve at present resonances of various multiplicities. Now near *any* resonance, stability increases, so that kind of mechanism can never reach the global exponent. Concretely, coming back to (∗), with $\mu = \varepsilon^2$, say, since we are working near a simple resonance, the stability exponent is at least $1/(2d)$ whereas we are in a system with $\ell = d + 1$ degrees of freedom. This may sound like a minor discrepancy but may also point to the fact that there exist other instability mechanisms which occur "far" from resonant surfaces and permit a larger speed of drift. These seem to be unknown at present and we note that the variational methods have recently come across the same problem; see the introduction of [Be2] where this point is briefly discussed.

We take this opportunity to just mention a few references dealing with the speed of *instability*. The general geometric strategy of this Arnold mechanism goes back to [A1] of course. The quantitative viewpoint was refined in [L2] and one can also consult [L4]. Since then [Mar1,2] and [Cr1,2] as well as other as yet unpublished works of these authors have improved and deepened the geometric features of the process, building partly on work of R.Easton (see the precise references in the aforementioned papers). These results use geometric constructions which in principle provide the possibility of reaching instability times which are nearly optimal (*i.e.* essentially match the stability theorems as far as exponents are concerned). Yet at the moment this still hangs on the possibility of getting bounds from below for the splitting, a matter which is discussed in Chapter 3 (see in particular §3.5) and seems to be quite hard for $\ell > 3$. However these works also suggest several possible really different new approaches.

A nice (mildly) variational approach was developed by U.Bessi (see [Be1,2,4]) which turned out to be quite efficient, at least for specific Hamiltonians because it requires rather precise system-dependent computations. In particular the case of the system originally introduced by Arnold in [A1] is studied in [Be1]. The method does not confront the singular perturbation problem so that it does not apply for polynomial (with respect to ε) values of μ in the notation of $(*)$ (see (1.4) and (2.1) in [Be1]) but it provides essentially optimal results in certain specific systems and for exponentially small μ (see [Be4]).

The simpler "*a priori* unstable" case ($\varepsilon = 1$, μ small) was studied in [CG] but the direct geometric method used there yields superexponentially long times (see in particular [CG] §8). Still, this was the first quantitative result in that field, and this paper sparked an enormous amount of additional work on this and related problems. Perhaps the most important point here (see [L4]) is that instability in that case is actually much stronger—and incidentally should therefore definitely *not* go by the name "Arnold diffusion". In fact the instability time is *polynomially* long, as explained in [L4] and shown in [B] on the example of the *a priori* unstable version of Arnold's original example, by adapting the method of [Be1]. Geometric methods as developped in [Mar1,2] and [Cr1,2] can now reach such results in more general cases, although much remains to be written up and further investigated.

2.6.6. We may now finally try to integrate the results of the present chapter into the incomplete picture we have sketched. As a matter of fact, the story begins with a prediction for the size of the splitting, as formalized in [L1] (reproduced in [L2] §V.2). It is this formal computation which predicts the equality of the four exponents mentioned in the introduction of this section and we will return to this point in §3.5, when discussing the connection between linear and nonlinear splitting. As for now we will see how the results of this chapter, embodied especially in the general results of §§2.4.8–2.4.9 fit into the landscape, which by now should be fairly natural and indeed almost obvious.

First let us quickly tour the respective settings again. In both cases, namely exponentially long stability time and exponentially small splitting, we need analytic data, and essentially for the same reason. However in the case of stability, one can decouple analysis from arithmetic *via* simultaneous approximation, which is not the case here for the splitting problem: we have used the Normal Form Lemma as initiated by Nekhoroshev. Restricting to the fully nonlinear case for simplicity, we come across an important difference: namely exponentially long stability requires convexity (or steepness), whereas exponentially small splitting does not. Concretely for $(*)$, as explained already in the general introduction, stability requires that all the components of α be of the same sign. Exponential smallness of the splitting does not. To be accurate we have seen that convexity *across* the resonance comes in handy ($B > 0$ in §§2.2.1–2.2.2) for ensuring the existence of certain objects, but is not absolutely necessary in general. It is also an empty condition for simple resonances. Below for the sake of comparison we have to assume that the unperturbed part is convex (with all sorts of possible weakenings available of course).

We now consider our favourite general near-integrable system (1) and localize to within $O(\sqrt{\varepsilon})$ near a resonant surface $S_\mathcal{M}$ of dimension n and multiplicity m, so associated to a lattice $\mathcal{M} \subset \mathbb{Z}^\ell$ of rank m. Perturbation theory predicts as we have seen an exponentially long stability time with exponent at least $1/(2n)$, and this is conjecturally generically optimal. We point out again that in order to speak of splitting at all we need a resonance to start with, so that $m > 0$, $n < \ell$ and $1/(2n) > 1/(2\ell)$ which is the predicted global stability exponent. This of course holds true all along $S_\mathcal{M}$, without any assumption of the initial local frequency being relatively Diophantine with respect to $\mathcal{M}$. All this is immediate when using simultaneous approximation. In the classical approach using small divisors (linear approximation), this is highly non-obvious and comes from the refined version of the Nekhoroshev puzzle as constructed by J.Pöschel. We refer to §2.1.5 above for some remarks on this, the point being that here in this chapter we have used only the analytic part (the Normal Form Lemma) of the classical approach but not the geometry of the puzzle.

So informally repeating the statement of Theorem 2.4.9 for convenience, we pick a relatively Diophantine point I_0 on the resonant surface with exponent $\tau_1(\geq n)$, and find that the corresponding splitting matrix has an n-dimensional spectral subspace which is ε-close to the tangent space of $S_\mathcal{M}$ at I_0. How does this fit with the picture provided by the local stability exponents? Well, homoclinic splitting is connected with stability and instability *via* Arnold mechanism, because just as in Arnold's original paper, homoclinic splitting is connected with *hetero*clinic splitting: by measuring the splitting of the invariant manifolds of *one* torus at an intersection point, or directly their distance away from intersections (which as we have seen already can be done by computing the difference of the *first* derivatives of the generating functions), one gets an idea of how far and how fast one can travel from one torus to another. The geometry developed in Chapter 1 should help in trying to unravel the situation. In any case passage from homoclinic to heteroclinic results already appears in [A1]. From the point of view of this chapter we point out again that if we want to study heteroclinic splitting, corresponding to *two* distinct tori, we need a patch in action space where we can build a common normal form. A main point is that *along* the resonance we can travel only exponentially slowly, so that we connect tori which are exponentially close and they do have a common normalization. So what we have done is (in principle!) sufficient to study the motion locally along a resonance.

Now *across* the resonance, first there is a "forbidden" direction coming from energy conservation in the (first-order) *averaged* system: all the splitting matrices we have seen have been computed after making a section, so they have size $(\ell - 1)$ and the missing direction is visible on the simplest of all, namely $M_0^\perp$ which is of size $m - 1$. It is the splitting matrix for the system governed by H_r^0 at the homoclinic point p (*cf.* §2.4.3) and we have discarded the direction along the homoclinic trajectory of this averaged system. So for simple resonances there remains *no* direction in which one can travel "polynomially fast" (or polynomially slowly if one prefers!). Now clearly, for $m > 1$, if one tries to "fool" the system and contradict the stability results using these polynomially small eigenvalues, that will fail because one will be able to travel polynomially fast over a distance on the order of $O(\sqrt{\varepsilon})$ away from the surface $S_\mathcal{M}$, but then the mechanism is broken and one has to consider other surfaces etc. Here the quantity $O(\sqrt{\varepsilon})$ refers to the resonance width and is to be compared with the trapping radius of the stability theorems, which is governed by a second exponent b; there are still some loose ends on this issue, which are discussed in [L2] and especially in [L3], §4.

But not all the pieces fit smoothly! There do remain puzzling features, even at this local (and heuristic!) level. We mention one of them as food for thought. With the notation as above, the splitting matrix has n eigenvalues with exponent $a = 1/(2\tau_1)$ and this analysis is in fact valid in a neighbourhood of size $O(\sqrt{\varepsilon})$ near I_0, in the sense that this is also the case for any torus in that neighbourhood (satisfying Assumptions 2.4.6, say). But if $\tau_1 > n$, the splitting exponent is thus *strictly smaller* than the local stability exponent, which seems to imply that one could try to contradict the stability results in a more subtle way, namely by traveling *along* the resonant surface, sliding from one relatively Diophantine torus to another, using tori which have Diophantine indices $> n$. In other words and very roughly speaking the splitting exponents match the local stability results only for tori which are as irrational as possible ($\tau_1 = n$; such vectors are called "badly approximable" in Diophantine approximation theory). But mathematics is mathematics and clearly this mechanism does not exist, because that would contradict the stability results, at least if traveling far enough (a distance larger than the trapping radius). This is somehow reminiscent of what we had a glimpse of in §2.6.4, namely the fully nonlinear theory hidden behind a relatively linear one. Perhaps this also shows that periodic orbits and simultaneous approximation should enter the game of the splitting and instability. More clues for this will appear below.

CHAPTER 3

The Hamilton-Jacobi Method For a Simple Resonance

We return in this section to the more specific setting attached to Hamiltonian $(*)$, namely again:

$$(*) \qquad H(q,\phi,p,I) = \omega \cdot I + \frac{1}{2}\alpha I^2 + \frac{1}{2}p^2 + \varepsilon(\cos q - 1) + \mu\varepsilon F.$$

It was designed to embody the main features of a near-integrable Hamiltonian in the vicinity of a simple resonance as illustrated in Paragraph 2.2 and we refer to the general introduction for a discussion of the meaning and relevance of the various parameters.

We develop in this chapter a method which is less robust and less geometric than the one exposed in the previous section, but which can reach more precise results under suitable hypotheses. Contrary to what happens in Chapter 2, it seemed it would have been too long to give all the details here and the reader is referred to [Sa2] for the complete proofs of the statements which appear below.

We will treat in detail the case of a perturbation F whose differential vanishes on a particular torus, thus leaving it invariant (see §3.1). In our opinion this help to isolate the mechanism which produces exponential smallness, since we can bypass KAM theory for finding an invariant torus; the general case is briefly discussed in §3.6.2. After reviewing the notation and assumptions in §3.1, we present in §3.2 a canonical scheme for computing the invariant manifolds at any given finite order; we call it the Hamilton-Jacobi algorithm, as it amounts to making full use of the Hamilton-Jacobi equation in the problem at hand. Perhaps the main results of this chapter are presented in §§3.3–3.4, where the exponential closeness of the invariant manifolds is proved by studying a certain vector field D on the configuration space. Then in §3.5 we recall the results and problems connected with evaluating the linear part of the splitting (the so-called Poincaré-Melnikov approximation); we are able, using the results of §3.4, to draw some precise conclusions in the case of three degrees of freedom. The next paragraph (§3.6) is devoted to discussing some variants and possible generalizations, without complete proofs.

Finally in §3.7, paralleling what we did at the end of Chapter 2, we give a historically based tour of the methods, results and problems on exponentially small splittings. We hope this will serve several purposes. The subject is quite technical and corresponds to just a fairly small portion of the picture we very partly delineated at the end of Chapter 2. But it is also certainly one of the most intensively studied problem in singular perturbation theory for differential equations, although we will not try to review the many connections: wave fronts, crystal growth etc. In Chapter 2 we introduced what we call the symplectic method, which to our knowledge had never been used before in the framework of the splitting problems. In the present chapter we use what we also call the *analytic* method, because analyticity seems to move to the front stage. Note that these should be seen only as temporary namesakes, given that analyticity is essential in order to make use of the symplectic method (and get *exponentially* small splittings) whereas the symplectic character of the problem is essential in our implementation of the analytic method. In fact we emphasized in the title the use of the Hamilton-Jacobi equation, which is one of the novel features here. We hope that we will have made these approaches look "natural" enough; §3.7 will also serve as a general conclusion for our study of the multidimensional splitting problems, especially in Chapters 2 and 3, and includes some general indications about why (not really 'how' unfortunately ...) the two methods should ultimately merge.

3.1. Notation and assumptions

3.1.1. In the whole of Chapter 3 we study the system governed by the Hamiltonian function (∗) recalled above. Let us briefly repeat for convenience some pieces of notation from the general introduction. We denote by ℓ ($\ell \geq 2$) the total number of degrees of freedom and write $\ell = d+1$ ($d \geq 1$), the conjugate variables being

$$(p, I) \in \mathbb{R} \times \mathbb{R}^d \quad \text{and} \quad (q, \phi) \in \mathbb{T} \times \mathbb{T}^d, \text{ with } \mathbb{T} = \mathbb{R}/2\pi\mathbb{Z}.$$

For completeness we mention that the phase space is naturally the cotangent bundle of $\mathbb{T} \times \mathbb{T}^d$, endowed with the exact symplectic structure induced by the Liouville form $\lambda = p\,dq + I \cdot d\phi$.

The parameters, whose meaning is discussed in the general introduction, are: a vector $\omega \in \mathbb{R}^d$; a real diagonal matrix $\alpha = \operatorname{diag}(\alpha_1, \ldots, \alpha_d)$ (the notation αI^2 means $\alpha I \cdot I = \sum_j \alpha_j I_j^2$ and we may also consider α as a d-vector); two small real parameters $\varepsilon > 0$ and μ; a real analytic function F.

3.1.2. The corresponding Hamiltonian system is integrable for $\mu = 0$, since it then decouples as the product of a simple pendulum and d independent rotators. We will refer to that situation as to the "unperturbed" one. In §§3.2–3.5, we impose specific assumptions on the perturbative term. These assumptions are partly simply for convenience as will be discussed further below, in particular in §3.6. So as for now, we make the following

STANDING ASSUMPTIONS (§§3.2–3.5).
i) The function F only depends on the angles (q, ϕ) and it extends analytically to $(\mathbb{C}/2\pi\mathbb{Z}) \times \mathbb{T}_{h_0}^d$ for some $h_0 > 0$, with the following notation:

$$\mathbb{T}_{h_0} = \{ \varphi \in \mathbb{C}/2\pi\mathbb{Z} \ ; \ |\Im m\,\varphi| < h_0 \}.$$

ii) For all $\phi \in \mathbb{T}^d$, $F(0, \phi) = 0$ and $\partial_q F(0, \phi) = 0$, *i.e.* F vanishes at order 2 on $\{q = 0\}$.

Taking Assumption i) into account, the equations of the motion read:

$$\begin{cases} \dot{q} = p \\ \dot{\phi}_j = \omega_j + \alpha_j I_j \\ \dot{p} = \varepsilon \sin q - \mu\varepsilon \partial_q F \\ \dot{I}_j = -\mu\varepsilon \partial_{\phi_j} F \end{cases}$$

The second assumption has an important effect: for each $J \in \mathbb{R}^d$, the Hamiltonian flow leaves the d-dimensional torus

$$\mathcal{T}_J = \{q = 0, \ \phi \in \mathbb{T}^d, \ p = 0, \ I = J\}$$

invariant, whatever the values of the parameters ε and μ (not even assumed to be small). Moreover the restriction of the motion to the torus $\mathcal{T}_J$ is quasiperiodic with quasifrequency $\omega + \alpha J$. Note that these tori are isotropic as it should be (see Proposition 1.8.4). The upshot is that in this chapter, due to this assumption and in contrast with the preceding one, we will not have to make use of any KAM technique (see §3.6.2 for more on this).

An important class of functions satisfying Assumption ii) is given by $F(q, \phi) = (\cos q - 1) f(q, \phi)$. The perturbative term then reads ε times $(\cos q - 1)(1 + \mu f(q, \phi))$ and one should assume that f satisfies Assumption i). One recovers in this way the class of systems which was considered in [L1], generalizing [A1]; see [L1] or [L2] (p. 117), where f was assumed to be independent of q for further simplification.

Finally without loss of generality we will restrict to the consideration of $J = 0$, which amounts to fixing the origin in I-space. The invariant torus $\mathcal{T}_0$ is 1-hyperbolic and we will study its unstable and stable invariant manifolds $\mathcal{W}^{\pm}$. Note that again we confine ourselves in this chapter to the study of a homoclinic problem.

3.1.3. Just as in Chapter 2 (see §2.2.1) it is convenient to rescale the time and the action variables: t is multiplied by $\varepsilon^{1/2}$ and the action variables p and I are divided by the same factor. This corresponds to Formulas (8) in §2.2.1 except that the situation was more general there, as we were studying a resonance of any multiplicity m. The variables (I, p) here correspond to (I_1, I_2) in Chapter 2, but with $m = 1$ so that $I_2 = p$ is a scalar variable.

So it is equivalent to study the Hamiltonian system generated by

$$(*) \qquad H_{\varepsilon,\mu}(q, \phi, p, I) = \varepsilon^{-1/2} \omega \cdot I + \frac{1}{2} \alpha I^2 + \frac{1}{2} p^2 + \cos q - 1 + \mu F(q, \phi),$$

with a large frequency vector $\varepsilon^{-1/2} \omega$ (compare §2.2.1). We still call this function $(*)$ and sometimes denote by z the variable $\varepsilon^{-1/2}$. All the statements below deal with that version of the Hamiltonian $(*)$.

3.1.4. Lastly two specific function spaces $\mathcal{B}^{\pm}$ will prove useful in the sequel.

NOTATION. We define $\mathcal{B}^-$ to be the space of all real analytic functions $U(q, \phi)$ which are defined for q close to 0 and $\phi \in \mathbb{T}^d$ and vanish at order 2 on $\{q = 0\}$; by this we mean that the function and all its first-order partial derivatives vanish on $\{q = 0\} \times \mathbb{T}^d$. Analogously we define $\mathcal{B}^+$ to be the space of all real analytic functions $U(q, \phi)$ which are defined for q close to 2π and $\phi \in \mathbb{T}^d$ and vanish at order 2 on $\{q = 2\pi\}$. Note that the function F belongs to both $\mathcal{B}^-$ and $\mathcal{B}^+$.

3.2. Formal solutions and the Hamilton-Jacobi algorithm

3.2.1. The invariant manifolds. For $\mu = 0$, $\mathcal{T}_0$ obviously admits stable and unstable manifolds which coincide and are given by the separatrix of the pendulum; we find it convenient to write them as:

$$\mathcal{W}^-_{|\mu=0} = \{ (q, \phi, p, I) \mid q \in]-2\pi, 2\pi[, \phi \in \mathbb{T}^d, p = 2 \sin \frac{q}{2}, I = 0 \},$$

$$\mathcal{W}^+_{|\mu=0} = \{ (q, \phi, p, I) \mid q \in]0, 4\pi[, \phi \in \mathbb{T}^d, p = 2 \sin \frac{q}{2}, I = 0 \},$$

distinguishing them arbitrarily only by their domains of definition. We should give different names to the tori $\{0\} \times \mathbb{T}^d \times \{0\} \times \{0\}$ and $\{2\pi\} \times \mathbb{T}^d \times \{0\} \times \{0\}$ as well, but we will refrain from doing so. Rather, from now on we will consider that the phase space is $\mathbb{R} \times \mathbb{T}^d \times \mathbb{R} \times \mathbb{R}^d$, which we identify with the cotangent bundle of the configuration space $\mathbb{R} \times \mathbb{T}^d$ where the variables (q, ϕ) live. Thus, above a point of the configuration space, covectors are identified with vectors of $\mathbb{R} \times \mathbb{R}^d$. Just as in Chapter 2 again, each of the unperturbed invariant manifolds is an exact Lagrangian graph, *i.e.* the graph of the differential of a function defined on part of $\mathbb{R} \times \mathbb{T}^d$:

$$\mathcal{W}^{\pm}_{|\mu=0} = \mathcal{G}r(dS_0) = \{\,(q,\phi,\partial_q S_0(q,\phi),\partial_\phi S_0(q,\phi))\,\}.$$

The generating function is given by:

$$S_0(q,\phi) = S_0(q) = 4(\cos\frac{q}{2} - 1),$$

and is considered as a function either on $]-2\pi, 2\pi[\times \mathbb{T}^d$ or on $]0, 4\pi[\times \mathbb{T}^d$.

As in Chapter 2 we will represent the perturbed invariant manifolds as graphs over this space; here we mean rather *parts* of the invariant manifolds which do not lie too far from the torus $\mathcal{T}_0$, *i.e.* local or "semi-local" stable and unstable manifolds. Another way to put it is that we use the compact-open topology, as in Chapter 2. So for nonzero μ we write:

$$\mathcal{W}^- = \mathcal{G}r(dS^-),\ \mathcal{W}^+ = \mathcal{G}r(dS^+),$$

where S^- and S^+ are functions on some parts of the configuration space which will be made precise in the statements below. These functions depend on the parameters ε and μ but here we keep this dependence implicit for notational simplicity. As in Chapter 2 once more, it is quite natural to do so since again the invariant manifolds, if they exist, have to be graphs (in the compact-open topology) because of their being close for small μ to $\mathcal{W}^{\pm}_{|\mu=0}$ which *is* a graph, and to be exact Lagrangian because of their being asymptotic to $\mathcal{T}_0$ which is an isotropic submanifold of the phase space (*cf.* §1.8.4).

3.2.2. To summarize, the setting is the same as in Chapter 2, only less general, and the notation is (almost) identical. In fact we diverge from Chapter 2 at this precise point, because instead of using symplectic resonant normal forms for trying to simplify the system itself, we now note that the generating functions $S^{\pm}$, and hence the manifolds $\mathcal{W}^{\pm}$, are determined up to an additive constant as the solutions of the Hamilton-Jacobi equation

(HJ) $$H_{\varepsilon,\mu}(q,\phi,\partial_q S(q,\phi),\partial_\phi S(q,\phi)) = 0,$$

such that both graphs $\mathcal{G}r(dS^{\pm})$ contain the torus $\mathcal{T}_0$. Note that here the right-hand side of (HJ) must vanish since $\mathcal{T}_0$ itself has zero energy.

The Hamilton-Jacobi equation is equivalent to the invariance of $\mathcal{G}r(dS^{\pm})$. The fact that $\mathcal{G}r(dS^-)$ (resp. $\mathcal{G}r(dS^+)$) must contain $\mathcal{T}_0$ amounts to the vanishing of the differential of $S^- - S_0$ (resp. $S^+ - S_0$) on $\{q = 0\}$ (resp. $\{q = 2\pi\}$); so the function $S^- - S_0$ (resp. $S^+ - S_0$) will be constant on $\{q = 0\}$ (resp. $\{q = 2\pi\}$) and these values, which here are not relevant, will be assumed to be zero, which offers a natural normalization for the generating functions $S^{\pm}$: we will simply require that $S^{\pm} - S_0 \in \mathcal{B}^{\pm}$ (using the notation introduced in §3.1.4).

The existence and regularity statement we need is embodied in the following

PROPOSITION. *For any $0 < q_0 < 2\pi$, there exists a positive constant μ_0 such that the Hamilton-Jacobi equation* (HJ) *admits a unique solution $S^-(q,\phi;\varepsilon,\mu)$*

which is real analytic in all its arguments for

$$-q_0 < q < q_0, \ \phi \in \mathbb{T}^d, \ \varepsilon > 0, \ |\mu| < \mu_0,$$

and such that $S^- - S_0$ belongs to $\mathcal{B}^-$ and vanishes identically when $\mu = 0$. Correspondingly (HJ) has a unique solution $S^+(q, \phi; \varepsilon, \mu)$ which is real analytic in all its arguments for

$$2\pi - q_0 < q < 2\pi + q_0, \ \phi \in \mathbb{T}^d, \ \varepsilon > 0, \ |\mu| < \mu_0,$$

and such that $S^+ - S_0$ belongs to $\mathcal{B}^+$ and vanishes identically when $\mu = 0$.

The invariant torus $\mathcal{T}_0$ then possesses stable and unstable manifolds which are locally the graphs of the differentials of the previous functions S^+ and S^- (differentials with respect to the variables q and ϕ):

$$\mathcal{W}^\pm = \mathcal{G}r(dS^\pm).$$

The critical points of the function defined by

$$\Delta S = S^+ - S^-, \ \text{for} \ 2\pi - q_0 < q < q_0 \ (q_0 > \pi), \ \phi \in \mathbb{T}^d, \ \varepsilon > 0, \ |\mu| < \mu_0,$$

are projections of intersections of $\mathcal{W}^-$ and $\mathcal{W}^+$.

Finally, the vector between $(Q, dS^-(Q))$ and $(Q, dS^+(Q))$ can be identified with $d(\Delta S)(Q)$ for each point $Q = (q, \phi)$.

This statement does not quite belong in this paragraph, where we will be especially busy with the *formal* side of the matter. We have inserted it partly in order to reassure the reader (and ourselves) as it says that we will actually be dealing with "real", not only formal objects. We will say a word about the proof of the first part of this proposition in §3.3 (the full proof appears in [Sa2]). The second and third parts are in fact obvious. They simply explain how the differential of ΔS is related to homoclinic trajectories and to the distance between $\mathcal{W}^-$ and $\mathcal{W}^+$.

The study of the splitting of the invariant manifolds is thus again reduced to the study of the differential of the function ΔS, much as in Chapter 2. The study of the homoclinic splitting at an intersection point is again encoded in the Hessian matrix of ΔS at this homoclinic point (see also §1.7.2). But the important point of difference between this and the last chapter consists roughly in the fact that in Chapter 2, after normalizing the system to a high order, we needed little regularity on the generating functions $S^\pm$ in order to derive exponentially small upper bounds. These functions simply had to be C^2 with respect to the space variables and we were not interested in the regularity with respect to the perturbation parameters. Here by contrast we will exploit *analyticity* in order to get bounds without any previous normalization of the system.

3.2.3. We now come to the main subject matter of this paragraph, as described in its title. Among other things Proposition 3.2.2 above claims the analyticity with respect to μ of the functions $S^\pm$. It is easy to obtain the corresponding Taylor expansions from the Hamilton-Jacobi equation (HJ): to this end it is sufficient to look for formal solutions of (HJ) of the type

$$S^\pm = S_0(q) + \sum_{n \geq 1} \mu^n S_n^\pm(q, \phi; \varepsilon),$$

where the coefficients $S_n^\pm$ are required to belong to $\mathcal{B}^\pm$.

Let us consider ε as fixed and write

$$T = \sum_{n \geq 0} \mu^n S_{n+1}(q, \phi)$$

for one of the two formal expansions that we are looking for. Clearly, $S = S_0 + \mu T$ is a solution of the Hamilton-Jacobi equation if and only if

$$D_0 T = -F(q, \phi) - \frac{1}{2}\mu[\alpha(\partial_\phi T)^2 + (\partial_q T)^2],$$

where we define

$$D_0 = \frac{dS_0}{dq}\frac{\partial}{\partial q} + \varepsilon^{-1/2}\omega \cdot \frac{\partial}{\partial \phi}$$

and call *unperturbed characteristic vector field* this vector field (or differential operator) on the configuration space.

We thus obtain the following set of recursive equations:

$$\begin{cases} D_0 S_1 = -F \\ D_0 S_{n+1} = -\dfrac{1}{2}\displaystyle\sum_{n_1+n_2=n-1}\left[\alpha \partial_\phi S_{n_1+1} \cdot \partial_\phi S_{n_2+1} + \partial_q S_{n_1+1}\partial_q S_{n_2+1}\right], & n \geq 1.\end{cases}$$

Since the coefficients $S_n = S_n^\pm$ we are interested in should lie in $\mathcal{B}^\pm$, in order to solve by induction the previous equations it is sufficient to invert D_0 on $\mathcal{B}^\pm$: the right-hand sides of these equations, beginning with F itself, will then belong to $\mathcal{B}^\pm$, as is easily checked by induction.

Now for $\mathcal{B}^-$, say, the reader may easily check the following

LEMMA. *The operator D_0 induces an automorphism of $\mathcal{B}^-$. In fact the change of variable $w = \tan\frac{q}{4}$ turns it into*

$$D_0 = w\frac{\partial}{\partial w} + \varepsilon^{-1/2}\omega \cdot \frac{\partial}{\partial \phi},$$

and this enables one to express the inverse $E^- = (D_{0|\mathcal{B}^-})^{-1}$ as

$$(E^- U)(w, \phi) = \int_{-\infty}^0 U(we^\zeta, \phi + \varepsilon^{-1/2}\zeta\omega)\, d\zeta.$$

Note that the integral above is absolutely convergent by the definition of $\mathcal{B}^-$. If U depends analytically on $z = \varepsilon^{-1/2}$ for $z > 0$, this is also the case for $E^- U$: indeed if z is allowed to move in a sector which contains $\mathbb{R}_+^*$, we can still change the half-line of integration $[0, -\infty[$ into $[0, -z^{-1}\infty[$ in order to keep $z\zeta$ real and the new formula will provide the analytic continuation of $E^- U$ with respect to z. Moreover, if $U(w, \phi)$ extends analytically to $\mathbb{R} \times \mathbb{T}^d$, which means that $U(\tan\frac{q}{4}, \phi)$ extends to $]-2\pi, 2\pi[\times\mathbb{T}^d$, as is the case for $F(q, \phi)$, the same thing is true of $E^- U$.

We can define analogously the inverse of $D_{0|\mathcal{B}^+}$ by using the change of variable $w' = \tan\frac{2\pi-q}{4}$. Denoting now by $E^\pm$ the inverse of $D_{0|\mathcal{B}^\pm}$ expressed in the original variables (q, ϕ), we summarize the algorithm we just obtained:

PROPOSITION. *For all $\varepsilon > 0$ there exist unique sequences $S_1^\pm, S_2^\pm, \ldots$ in $\mathcal{B}^\pm$ such that the series*

(1) $$S^\pm = S_0(q) + \sum_{n \geq 1} \mu^n S_n^\pm(q, \phi)$$

formally satisfy the Hamilton-Jacobi equation (HJ).

The functions S_n^- (resp. S_n^+) extend to $]-2\pi, 2\pi[\times \mathbb{T}^d$ (resp. $]0, 4\pi[\times \mathbb{T}^d$) and all of them depend analytically on $z = \varepsilon^{-1/2}$. They are determined by the following inductive formulas:

(2) $$\begin{cases} S_1^\pm = -E^\pm F \\ S_{n+1}^\pm = -\frac{1}{2} E^\pm \Big(\sum_{n_1 + n_2 = n-1} [\alpha \partial_\phi S_{n_1+1}^\pm \cdot \partial_\phi S_{n_2+1}^\pm + \partial_q S_{n_1+1}^\pm \partial_q S_{n_2+1}^\pm] \Big), \\ \hspace{8cm} n \geq 1, \end{cases}$$

and these series are the Taylor expansions in μ of the functions $S^\pm(q, \phi; \varepsilon, \mu)$ in Proposition 3.2.2.

3.2.4. Since our goal is to compare S^+ and S^-, we can restrict attention to the common domain of definition $]0, 2\pi[\times \mathbb{T}^d$ and it is interesting to have an expression of the operators E^+ and E^- involving the same variables. To this end we define the change of variable:

$$q = q_0(u) = 4 \arctan e^u.$$

The variable u, which was already used by Poincaré, is nothing but the time variable along the separatrix of the pendulum and it will prove very useful in the sequel. In particular it will be essential to see it as a complex variable, and to try and enlarge as much as possible the domains of analyticity with respect to u of various functions defined on configuration space. We will use a tilde over the symbol denoting such a function in order to indicate that we have performed the change of variable $q = q_0(u)$: $\tilde{F}(u, \phi) = F(q_0(u), \phi)$, etc.

The function $q_0(u)$ itself extends analytically to the universal covering of the cut plane $\mathbb{C} \setminus (\frac{i\pi}{2} + i\pi \mathbb{Z})$ with logarithmic singularities only, and it defines a (uniform) analytic $(2\pi i)$-periodic function on

$$\mathcal{C} = \mathbb{C} \setminus \big([\frac{i\pi}{2}, \frac{3i\pi}{2}] + 2i\pi \mathbb{Z}\big).$$

The image of $\mathcal{C}$ by q_0 is the vertical strip $\{q \in \mathbb{C} \mid -\pi < \Re q < 3\pi\}$, except for the points 0 and 2π which are obtained as limits when $\Re u$ tends to $-\infty$ or $+\infty$ respectively. The singular points $i\pi/2$ and $3i\pi/2$ correspond to $\Im q = +\infty$ and $\Im q = -\infty$ respectively, and the left and right sides of the cut $]\frac{i\pi}{2}, \frac{3i\pi}{2}[$ correspond to the vertical boundaries $\Re q = -\pi$ and $\Re q = 3\pi$ respectively. This elementary analysis of the conformal mapping $q(u)$ is given for pictorial completeness only.

Since $q_0(u)$ is analytic over $\mathcal{C}$, the function $\tilde{F}$ is analytic over $\mathcal{C} \times \mathbb{T}_{h_0}^d$. The counterpart of the unperturbed characteristic vector field D_0 reads

$$\tilde{D}_0 = \frac{\partial}{\partial u} + \varepsilon^{-1/2} \omega \cdot \frac{\partial}{\partial \phi},$$

and one easily computes the counterparts $\tilde{E}^\pm$ of the operators $E^\pm$; namely we have the following easy

LEMMA. *If $U \in \mathcal{B}^\pm$ and letting $\tilde{U}(u,\phi) = U(q_0(u),\phi)$,*

$$(E^\pm U)(q_0(u),\phi) = \tilde{E}^\pm \tilde{U}(u,\phi) = -\int_0^{\pm\infty} \tilde{U}(u+\zeta, \phi+\varepsilon^{-1/2}\zeta\omega)\,d\zeta.$$

The integral converges absolutely because $\tilde{U}(u,\phi) = O(e^{-2|\Re e\,u|})$ uniformly in $\phi \in \mathbb{T}^d$ when $\Re e\,u$ tends to $\pm\infty$. The point is that this lemma allows us to compute inductively and fairly easily the functions $\tilde{S}_n^\pm(u,\phi;\varepsilon)$ which correspond to the coefficients of (2) in the variables (u,ϕ).

3.2.5. In order to illustrate the above we spell out the first two orders of the algorithm. The first-order approximation of $S^\pm$ is $S_0 + \mu S_1^\pm$ where

$$S_1^\pm(q_0(u),\phi;\varepsilon) = \tilde{S}_1^\pm(u,\phi;\varepsilon) = \int_0^{\pm\infty} \tilde{F}(u+\zeta, \phi+\varepsilon^{-1/2}\zeta\omega)\,d\zeta.$$

Not surprisingly, the difference $\Delta S_1 = S_1^+ - S_1^-$ is the familiar Poincaré-Melnikov integral:

$$(3) \qquad \Delta \tilde{S}_1(u,\phi;\varepsilon) = \int_{-\infty}^{+\infty} \tilde{F}(u+\zeta, \phi+\varepsilon^{-1/2}\zeta\omega)\,d\zeta.$$

Now for $n=2$ the Hamilton-Jacobi algorithm yields

$$\tilde{S}_2^\pm = -\frac{1}{2}\tilde{E}^\pm \big[\alpha(\partial_\phi \tilde{S}_1^\pm)^2 + \beta^2(\partial_u \tilde{S}_1^\pm)^2\big],$$

where the function $\beta(u) = \frac{1}{2}\cosh u = (\frac{dq_0}{du})^{-1}$ stems from the change of variable $q = q_0(u)$. At the price of a slight manipulation of integrals, the reader may check that

$$\tilde{S}_2^\pm(u,\phi;\varepsilon) = \int_0^{\pm\infty} \hat{S}_2(u,\phi,\zeta)\,d\zeta,$$

with $\hat{S}_2(u,\phi,\zeta)$ defined as the sum

$$\alpha \partial_\phi \tilde{F}(u+\zeta, \phi+\varepsilon^{-1/2}\zeta\omega) \cdot \int_0^\zeta \zeta' \partial_\phi \tilde{F}(u+\zeta', \phi+\varepsilon^{-1/2}\zeta'\omega)\,d\zeta'$$

$$+ \partial_u \tilde{F}(u+\zeta, \phi+\varepsilon^{-1/2}\zeta\omega) \int_0^\zeta (B(u+\zeta')-B(u))\partial_u \tilde{F}(u+\zeta', \phi+\varepsilon^{-1/2}\zeta'\omega)\,d\zeta',$$

where $B(u) = \int_0^u \beta^2(u')du' = \frac{1}{8}(u + \frac{\sinh 2u}{2})$. This gives a fairly simple expression for the difference $\Delta S_2 = S_2^+ - S_2^-$ in the variable u:

$$(4) \qquad \Delta \tilde{S}_2(u,\phi;\varepsilon) = \int_{-\infty}^{+\infty} \hat{S}_2(u,\phi,\zeta)\,d\zeta.$$

3.2.6. Before moving from the formal to the analytic problems, we would like to make two short remarks about the above. The first is that one can easily write down an algorithm which generalizes the one given here in Proposition 3.2.3 to the case where F does depend on the action variables (p,I); this will be developed in §3.6.1. Formally it is just a little more cumbersome, and at this formal level Assumption 3.1.2 i) saying that F depends on the angles only has been imposed only for sake of simplicity of the exposition. On the other hand, if one sticks to Assumption 3.1.2 i) and assumes furthermore that $\alpha = 0$, *i.e.* if one is interested in the torsionfree (or isochronous) case, the algorithm very substantially simplifies. As is clear from the inductive formulas (2), the angular derivatives of the coefficients disappear and this is a serious simplification, both at the practical computational level and at the theoretical level as will be shown below in §3.3.4. Clearly the two situations can mix, that is one can consider an algorithm for the isochronous case but with an action-dependent perturbation.

The second remark, which we will expand a bit in the concluding Paragraph 3.7, is that it seems to us that the algorithm given above is the "right" one if one wishes to compute or estimate the invariant manifolds and the splitting matrix by direct term-by-term evaluation. In other words it offers a solid basis for the development of "direct methods" in this context. Let us confine ourselves to mentioning some obvious facts. Here we only need to compute two *scalar* functions $S^\pm$ and their difference. Moreover the computations for S^+ and S^- are in fact completely parallel, the changes basically amounting to switching certain signs (we did not make this last operation completely algorithmic though). At least we have reduced to the minimum the amount of relevant information: in this Hamiltonian case, it is *indeed* carried by two scalar functions $S^\pm$, or even just one (namely the difference $\Delta S = S^+ - S^-$) if one is interested in the distance between the invariant manifolds and the splitting matrices at the intersection points. This has already been amply illustrated in Chapters 1 and 2. We insist on this point, at the risk of rubbing it in too heavily as it may indeed appear "obvious" in this context (see also §3.7.3). But another fact is that the algorithms which have been used hitherto do *not* fully take advantage of the symplectic character of the problem. They compute far too many quantities which carry the *same* information as here, only in a very redundant way. Since the symmetries of the problem are not built into the algorithm, they reappear in the form of seemingly "mysterious" cancellations which are often very difficult to trace. Here again and by contrast, the information is conveyed in a minimal way: this could be expressed in a statement saying that the *only* algebraic or differential relations connecting the coefficients $S_n^\pm$ are generically those given by the algorithm itself and "trivial" consequences thereof; in other words Equations (2), viewed as relations among the $S_n^\pm$, should generate the differential ideal of the (generic) relations.

3.3. Convergence and domains of analyticity

3.3.1. So far we have not seen how to prove Proposition 3.2.2, *i.e.* how to show the convergence of the series (1). A natural approach would be to rely on the previous algorithm and to bound inductively the coefficients $S_n^\pm$ in some suitable Banach space. But in view of the induction formulas (2), this would require to bound the operators $\partial_q \circ E^\pm$ and $\partial_{\phi_j} \circ E^\pm$, rather than the integral operator $E^\pm$ itself.

It turns out that we can define a Banach norm for which $E^\pm$ and $\partial_q \circ E^\pm$ are bounded, but we do not know how to do the same for $\partial_{\phi_j} \circ E^\pm$ except when $d = 1$. Of course the difficulty disappears if $\alpha = 0$, so that there is a shortcut (compared to what we present below in §3.3.5) both in the one-frequency case and in the multifrequency anisochronous case. We note that to the best of our knowledge these two cases cover all the existing literature to-date, except for [RW1] which contains errors, hopefully to be corrected (see the erratum to that paper and [RW3]). We devote §3.3.4 to the case where $\alpha = 0$, with some indications for the case where $d = 1$.

In order to deal with the general case, *i.e.* for an arbitrary value of the twist vector α, we had to devise an indirect method. It is in fact the only place where we depart from the symplectic framework, using more hyperbolic tools instead. In other words we do *not* at this point exclusively make use of the Hamilton-Jacobi equation and do not know whether that would be feasible. We will state a theorem which is stronger than Proposition 3.2.2 and which deals with the counterpart in the variables (u, ϕ) of the series (1), that is:

$$(5) \qquad \tilde{S}^\pm = \tilde{S}_0(u) + \sum_{n \geq 1} \mu^n \tilde{S}_n^\pm(u, \phi) = S_0(q_0(u)) + \sum_{n \geq 1} \mu^n S_n^\pm(q_0(u), \phi).$$

Returning to the variables (q, ϕ) is then easy. We will need to first define a few quantities that are required for the statement of the result, then we give the statement (Theorem 3.3.3), the complete proof in the simpler case where $\alpha = 0$ (§3.3.4) and some indications about the proof method in the general case (§3.3.5—the complete proof is contained in [Sa2], §5).

3.3.2. The function $A(\delta, \sigma)$. We have seen that $\tilde{F}$ is analytic for $u \in \mathcal{C} = \mathbb{C} \setminus ([\frac{i\pi}{2}, \frac{3i\pi}{2}] + 2i\pi\mathbb{Z})$ and $\phi \in \mathbb{T}^d$. For $0 < \delta < \pi/2$, we will denote by $\mathcal{C}_\delta$ a subset of $\mathcal{C}$ which contains $\mathbb{R}$:

$$\mathcal{C}_\delta = \{ u \in \mathbb{C} \mid \text{dist}(u, [\frac{i\pi}{2}, \frac{3i\pi}{2}] + 2i\pi\mathbb{Z}) \geq \delta \}.$$

For all $\delta, \sigma > 0$ (with $\delta < \pi/2$ and $\sigma < h_0$), there exists a number $A = A(\delta, \sigma) \geq 1$ such that

$$\forall (u, \phi) \in \mathcal{C}_\delta \times \overline{\mathbb{T}}^d_{h_0 - \sigma}, \quad |\tilde{F}(u, \phi)| \leq A\, e^{-2|\Re e\, u|},$$

where we have used the notation $\overline{\mathbb{T}}_h = \{ \phi \in \mathbb{C}/2\pi\mathbb{Z};\; |\Im m\, \phi| \leq h \}$ if $h > 0$. We will consider this function $A(.,.)$ as a datum of the problem in the same way as the function F itself; it is in fact a way of measuring the size of F, or the strength of the singularities of $\tilde{F}$ on the imaginary axis for the variable u and on the boundary of $\overline{\mathbb{T}}^d_{h_0}$ for the variable ϕ. One may keep in mind the typical example of a function like $A(\delta, \sigma) = \text{cst}\, \delta^{-n} \sigma^{-m}$, with $n, m \in \mathbb{Z}_+^*$, which corresponds to polar singularities.

What matters here is the size of the perturbative term F along the complexified unperturbed manifold $\mathcal{W}^\pm_{|\mu=0}$. Were F to depend also on the action variables (p, I), we would have to replace the bound on $\tilde{F}(u, \phi)$ by one on $F(q_0(u), \phi, \dot{q}_0(u), 0)$.

3.3.3. The domains of analyticity for $\tilde{S}^\pm$. We are now going to state the central analytical result of this chapter. First we have to define the complex domains $\mathcal{D}^-_{u_1, \delta}$ and $\mathcal{D}^+_{u_1, \delta}$ in the u-plane, over which the series (5) will be proved to converge. For $u_1 \in \mathbb{R}$ and $0 < \delta < \pi/2$, we consider all the open sectors

$$\{ u = u_1 - \xi\, e^{i\beta};\; \xi > 0,\; \beta \in]-\beta_1, \beta_1[\}$$

for $\beta_1 \in]0, \pi/2]$, and select among them the largest which is contained in $\mathcal{C}_\delta$: we call $\mathcal{D}^-_{u_1, \delta}$ this sector. If $u_1 > -\delta$, its vertex is u_1 and its aperture $2\beta_1$ is determined by the equation

$$\frac{\pi}{2} \cos \beta_1 = \delta + u_1 \sin \beta_1,$$

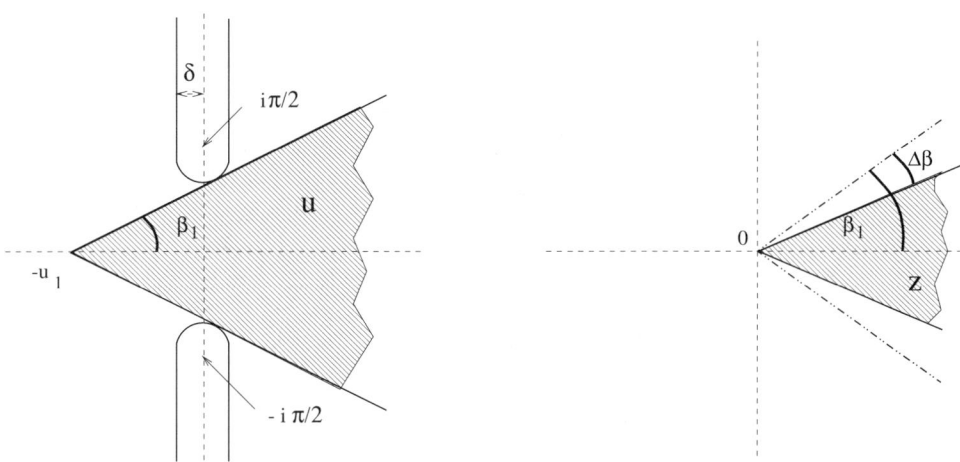

FIGURE 1. The domain of analyticity $\mathcal{D}^+_{u_1,\delta} \times \Sigma_{u_1,\delta,\Delta\beta}$ in the variables (u,z).

but if $u_1 \leq -\delta$, it is the open half-plane $\{\Re e\, u < u_1\}$ and $\beta_1 = \pi/2$.

Similarly, $\mathcal{D}^+_{u_1,\delta}$ is the largest of the sectors

$$\{\, u = -u_1 + \xi e^{i\beta};\ \xi > 0,\ \beta \in]-\beta_1,\beta_1[\,\}, \qquad \beta_1 \in]0,\pi/2],$$

which is contained in $\mathcal{C}_\delta$; it is in fact just the opposite sector:

$$\mathcal{D}^+_{u_1,\delta} = \{\, -u;\ u \in \mathcal{D}^-_{u_1,\delta}\,\}.$$

Lastly we define a complex domain for the variable $z = \varepsilon^{-1/2}$: for $\Delta\beta \in]0,\pi/2[$, with moreover $\Delta\beta < \arctan \frac{\pi}{2u_1}$ if $u_1 > 0$,

$$\Sigma_{u_1,\delta,\Delta\beta} := \{\, z = \xi e^{i\beta};\ \xi > 0,\ \beta \in]-\beta_1 + \Delta\beta, \beta_1 - \Delta\beta[\,\},$$

where the half-aperture $\beta_1 \in]0,\pi/2]$ of the sectors $\mathcal{D}^-_{u_1,\delta}$ and $\mathcal{D}^+_{u_1,\delta}$ is supposed to be strictly larger than $\Delta\beta$; this will be the case if δ is small enough with respect to u_1 and $\Delta\beta$. (See Figure 1.)

We may now state the general analyticity result for the functions $\tilde{S}^\pm$ which correspond to the solutions of (HJ) we are interested in:

THEOREM. *For all $u_1 \in \mathbb{R}$ and $\Delta\beta \in]0,\pi/2[$ with $\Delta\beta < \arctan \frac{\pi}{2u_1}$ if $u_1 > 0$, and for all small enough $\delta, \sigma > 0$, there exist positive numbers μ_1 and $\{B_\partial\}$ such that the series (5) converge to functions $\tilde{S}^\pm(u,\phi;z,\mu)$ which are analytic with respect to all their arguments for*

$$u \in \mathcal{D}^\pm_{u_1,\delta},\ \phi \in \mathbb{T}^d_{h_0-\sigma},\ z = \varepsilon^{-1/2} \in \Sigma_{u_1,\delta,\Delta\beta},\ |\mu| < \mu_1,$$

and which satisfy the following inequalities on the closure of these domains:

$$|\partial(\tilde{S}^\pm - \tilde{S}_0 - \mu \tilde{S}_1^\pm)| \leq B_\partial\, |\mu|^2\, e^{-(\pm 2\Re e\, u)},$$

where ∂ stands for one of the operators $(\partial_u)^{r_0}(\partial_\phi)^r$ with $(r_0,r) \in \mathbb{Z}_+ \times \mathbb{Z}_+^d$ such that $r_0 + |r| \leq 2$. More precisely, if $\delta \leq \sigma$, one can choose

$$\mu_1 = b_1^{-1}\, A\, (\delta/2, \sigma/2)^{-1}\, \delta\, \sigma^{d+1}$$

and
$$B_\partial = \begin{cases} b_1 \, A \, (\delta/2, \sigma/2)^2 \, \delta^{-2-\sup\{1,r_0\}} \sigma^{-2d-1} & \text{if } |r| = 0, \\ b_1 \, A \, (\delta/2, \sigma/2)^2 \, \delta^{-2} \sigma^{-2d-|r|} & \text{if } |r| \geq 1, \end{cases}$$

where the positive number b_1 depends only on u_1 and $\Delta\beta$.

Of course here $(\partial_\phi)^r$ means $(\partial_{\phi_1})^{r_1} \ldots (\partial_{\phi_d})^{r_d}$ and $|r| = r_1 + \cdots + r_d$.

3.3.4. Proof of Theorem 3.3.3 in the isochronous case ($\alpha = 0$). In fact we will obtain better estimates in that case; we provide the details since they are not given in [Sa2]. Let us for instance consider the case of the stable manifold. We fix u_1 and $\Delta\beta$ as in the statement of the theorem, and $\delta, \sigma \in \,]0, 1[$. We will prove the uniform convergence of the series (5) (with a 'plus' sign) over $\overline{\mathcal{D}}^+_{u_1, \frac{3\delta}{4}} \times \overline{\mathbb{T}}^d_{h_0 - \frac{3\sigma}{4}}$ for $z \in \overline{\Sigma}_{u_1, \frac{3\delta}{4}, \Delta\beta}$ and $|\mu|$ small enough, by bounding inductively the size of each coefficient, and then use the Cauchy inequalities in the smaller domain $\overline{\mathcal{D}}_{u_1, \delta} \times \overline{\mathbb{T}}^d_{h_0 - \sigma}$. We recall the induction formulas in that case:

$$\begin{cases} \tilde{S}^+_1 = -\tilde{E}^+ \tilde{F} \\ \tilde{S}^+_{n+1} = -\frac{1}{2}\tilde{E}^+ \Big(\sum_{n_1+n_2=n-1} \beta^2 \partial_u \tilde{S}^+_{n_1+1} \partial_u \tilde{S}^+_{n_2+1} \Big), \quad n \geq 1, \end{cases}$$

with $\beta(u) = \frac{1}{2} \cosh u$. We begin by defining the appropriate Banach algebra, which involves "Fourier norms".

We denote by $\mathbb{B}$ the space spanned by the functions $\tilde{U}$ which are analytic in $\mathcal{D}^+_{u_1, \frac{3\delta}{4}} \times \Sigma_{u_1, \frac{3\delta}{4}, \Delta\beta}$, extend continuously to $\overline{\mathcal{D}}^+_{u_1, \frac{3\delta}{4}} \times \overline{\Sigma}_{u_1, \frac{3\delta}{4}, \Delta\beta}$ and satisfy

$$\|\tilde{U}\| := \sup_{(u,z) \in \overline{\mathcal{D}}^+_{u_1, \frac{3\delta}{4}} \times \overline{\Sigma}_{u_1, \frac{3\delta}{4}, \Delta\beta}} |e^{2u_1 + 2\Re e \, u} \, \tilde{U}(u,z)| < +\infty.$$

Since $2u_1 + 2\Re e \, u \geq 0$ for all u in $\overline{\mathcal{D}}^+_{u_1, \frac{3\delta}{4}}$, this defines a Banach algebra $(\mathbb{B}, \|\cdot\|)$. By requiring that the Fourier coefficients of our functions belong to it, we will ensure their exponential decay at infinity.

For any $h > 0$, we denote by $\mathbb{B}(h)$ the space consisting of all the Fourier series

$$\tilde{U} = \sum_{k \in \mathbb{Z}^d} e^{ik\cdot\phi} \tilde{U}_k \qquad \text{with } (\forall k \in \mathbb{Z}^d) \, \tilde{U}_k \in \mathbb{B},$$

such that

$$\|\tilde{U}\|_h := \sum_{k \in \mathbb{Z}^d} e^{|k|h} \|\tilde{U}_k\| < +\infty,$$

where $k \cdot \phi = k_1 \phi_1 + \cdots + k_d \phi_d$ and $|k| = |k_1| + \cdots + |k_d|$. This defines a Banach algebra $(\mathbb{B}(h), \|\cdot\|_h)$. Any Fourier series $\tilde{U}$ in $\mathbb{B}(h)$ converges to an analytic mapping $\tilde{U}$ from $\mathbb{T}^d_h$ to $\mathbb{B}$ which extends continuously to $\overline{\mathbb{T}}^d_h$ and satisfies $\sup_{\overline{\mathbb{T}}^d_h} \|\tilde{U}\| \leq \|\tilde{U}\|_h$. Moreover, if $0 < h' < h$, there is another relation between its Fourier norm and its sup norm:

$$\|\tilde{U}\|_{h'} \leq \coth^d\big(\tfrac{h-h'}{2}\big) \sup_{\overline{\mathbb{T}}^d_h} \|\tilde{U}\|,$$

and the Cauchy inequalities hold

$$\forall r \in \mathbb{Z}^d_+, \quad \|(\partial_\phi)^r \tilde{U}\|_{h'} \leq r_1! \ldots r_d! (h-h')^{-|r|} \|\tilde{U}\|_h.$$

Such a construction was already used in [Pö1] and [RW1]. The reason why we use this kind of norm for Fourier series appears in the

LEMMA. *Let $h > 0$. The operators $\tilde{E}^+$ and $\partial_u \circ \tilde{E}^+$ induce bounded operators of $\mathbb{B}(h)$ whose operator norm is less or equal to $\kappa = 2/\sin\Delta\beta$.*

PROOF. Let $\tilde{U} \in \mathbb{B}(h)$. We first check that $\tilde{V} = \tilde{E}^+ \tilde{U}$ belongs to $\mathbb{B}(h)$ by writing it as
$$\tilde{V}(u, \phi, z) = -\int_0^{z^{-1}\infty} \tilde{U}(u + \zeta, \phi + z\zeta\omega, z)\, d\zeta,$$
that is
$$\forall (u, z) \in \overline{\mathcal{D}}^+_{u_1, \frac{3\delta}{4}} \times \overline{\Sigma}_{u_1, \frac{3\delta}{4}, \Delta\beta}, \quad \tilde{V}_k(u, z) = -\int_0^{z^{-1}\infty} \tilde{U}_k(u + \zeta, z)\, e^{iz\zeta k \cdot \omega}\, d\zeta.$$
(Observe that $|\arg \zeta| = |\arg z^{-1}| \leq \beta_1$, the half-aperture of the sector $\mathcal{D}^+_{u_1, \frac{3\delta}{4}}$, thus $u + \zeta \in \overline{\mathcal{D}}^+_{u_1, \frac{3\delta}{4}}$). This integral converges since
$$|\tilde{U}_k(u + \zeta, z)| \leq e^{-2u_1 - 2\Re e\, u}\|\tilde{U}_k\|\, e^{-2\Re e\, \zeta}$$
and $\Re e\, \zeta = |\zeta|\cos(\arg z)$ with $\cos(\arg z) \geq \sin\Delta\beta$ (because $|\arg z| \leq \frac{\pi}{2} - \Delta\beta$). Moreover the inequalities $|\tilde{V}_k(u, z)| \leq e^{-2u_1 - 2\Re e\, u}\|\psi_k\|(2\sin\Delta\beta)^{-1}$ show that $\tilde{V}$ belongs to $\mathbb{B}(h)$ and
$$\|\tilde{V}\|_h \leq \frac{\kappa}{2}\|\tilde{U}\|_h.$$

Let $(u, z) \in \overline{\mathcal{D}}^+_{u_1, \frac{3\delta}{4}} \times \overline{\Sigma}_{u_1, \frac{3\delta}{4}, \Delta\beta}$. The Cauchy theorem allows us to move the half-line of integration in the definition of $\tilde{V}_k$. If $k \cdot \omega > 0$, we increase the slope of the half-line in order to take advantage of the decrease of the exponential $e^{iz\zeta k \cdot \omega}$:
$$\tilde{V}_k(u, z) = -\int_0^{+\infty} \tilde{U}_k(u + \xi e^{i\beta_1}, z)\, e^{ize^{i\beta_1}\xi k \cdot \omega}\, e^{i\beta_1}\, d\xi,$$
and we obtain a new bound:
$$\|\tilde{V}_k\| \leq \frac{1}{2\cos\beta_1 + |z|k \cdot \omega \sin(\beta_1 + \arg z)}\|\tilde{U}_k\|.$$
Since $0 < \beta_1 \leq \frac{\pi}{2}$ and $\Delta\beta \leq \beta_1 + \arg z \leq \pi - \Delta\beta$, we end up with
$$\|zk \cdot \omega \tilde{V}_k\| \leq \frac{\kappa}{2}\|\tilde{U}_k\|.$$
Similarly, if $k \cdot \omega < 0$, we decrease the slope of the half-line of integration, and the previous inequality holds true in all cases (even if $k \cdot \omega = 0$). This means that
$$\|z\omega \cdot \partial_\phi V\|_h \leq \frac{\kappa}{2}\|\tilde{U}\|_h.$$
But $\partial_u \tilde{V} = \tilde{U} - z\omega \cdot \partial_\phi \tilde{V}$ because $\tilde{D}_0 \tilde{V} = \tilde{U}$; thus
$$\|\partial_u \tilde{V}\|_h \leq \kappa \|\tilde{U}\|_h$$
and we are done. The above proof "explains" why we had to define such a domain of analyticity as the sector $\mathcal{D}^+_{u_1, \frac{3\delta}{4}}$. □

Let $\lambda = \max\{1, \frac{1}{2}\cosh u_1\}$: we leave it to the reader to check that, since $|\beta(u)| \leq \lambda e^{u_1 + \Re e\, u}$ in $\overline{\mathcal{D}}^+_{u_1, \frac{3\delta}{4}}$,

$$\forall h > 0, \ \forall \tilde{U}, \tilde{V} \in \mathbb{B}(h), \qquad \beta^2 \tilde{U}\tilde{V} \in \mathbb{B}(h) \ \text{and} \ \|\beta^2 \tilde{U}\tilde{V}\|_h \leq \lambda^2 \|\tilde{U}\|_h \|\tilde{V}\|_h.$$

We will use $h = h_0 - \frac{3\sigma}{4}$ in the sequel. Let

$$A = \sigma^{-d} A(\tfrac{\delta}{2}, \tfrac{\sigma}{2}).$$

We observe that $\tilde{F} \in \mathbb{B}(h)$ and $\|\tilde{F}\|_h \leq bA$, with a positive number b which depends only on u_1 (using the inequality above between Fourier and sup norms). We are now in a position to check that the functions $\tilde{S}_n^+$ and $\partial_u \tilde{S}_n^+$ belongs to $\mathbb{B}(h)$, and to define a sequence $(\mathcal{S}_n)_{n \geq 1}$ of positive numbers such that

$$\forall n \geq 1, \qquad \|\tilde{S}_n^+\|_h, \ \|\partial_u \tilde{S}_n^+\|_h \leq \mathcal{S}_n.$$

It suffices indeed to choose $\mathcal{S}_1 = \kappa bA$ and

$$\forall n \geq 1, \quad \mathcal{S}_{n+1} = \frac{1}{2}\kappa \sum_{n_1 + n_2 = n-1} \lambda^2 \mathcal{S}_{n_1} \mathcal{S}_{n_2}.$$

The convergence of the generating series $\mathcal{S}(\mu) = \sum_{n \geq 0} \mu^n \mathcal{S}_{n+1}$ is easily studied: The inductive definition of the numbers $\mathcal{S}_n$ amounts to the equation

$$\mathcal{S} = \mathcal{S}_1 + \frac{\kappa \lambda^2}{2}\mu \mathcal{S}^2,$$

whence the formula $\mathcal{S} = \mathcal{S}_1 R(\kappa \lambda^2 \mu \mathcal{S}_1)$ with $R(X) = \frac{1 - (1-2X)^{1/2}}{X}$. We retain that

$$\forall n \geq 0, \qquad \mathcal{S}_{n+1} = R_n (\kappa \lambda^2)^n \mathcal{S}_1^{n+1} \ \text{with} \ R_n \leq \text{cst}\, 2^{-n}.$$

We end up with inequalities

$$\forall n \geq 1, \qquad \|\tilde{S}_n^+\|_h, \ \|\partial_u \tilde{S}_n^+\|_h \leq \text{cst}\, \Big(\frac{\kappa \lambda^2}{2}\Big)^n \mathcal{S}_1^n \ \text{with} \ \mathcal{S}_1 = \kappa bA.$$

This ensures the convergence of $\tilde{S}^+$ for

$$|\mu| \leq \mu_1 = b_1^{-1} \sigma^d A(\tfrac{\delta}{2}, \tfrac{\sigma}{2})^{-1}$$

with $b_1 = \kappa^2 \lambda^2 b$. Note that this value of μ_1 is better than the one announced in Theorem 3.3.3. In that range, we have (enlarging b_1 if necessary)

$$\|\tilde{S}^+ - \tilde{S}_0 - \mu \tilde{S}_1^+\|_{h_0 - \frac{3\sigma}{4}}, \ \|\partial_u(\tilde{S}^+ - \tilde{S}_0 - \mu \tilde{S}_1^+)\|_{h_0 - \frac{3\sigma}{4}} \leq b_1 A(\tfrac{\delta}{2}, \tfrac{\sigma}{2})^2 \sigma^{-2d}.$$

We conclude using Cauchy inequalities (w.r.t. u and ϕ) which yield bounds relative to $\overline{\mathcal{D}}^+_{u_1, \delta}$ and $\overline{\mathbb{T}}^d_{h_0 - \sigma}$:

$$|(\partial_u)^{r_0}(\partial_\phi)^r (\tilde{S}^+ - \tilde{S}_0 - \mu \tilde{S}_1^+)| \leq \text{cst}\, A(\tfrac{\delta}{2}, \tfrac{\sigma}{2})^2 \delta^{1 - \sup\{1, r_0\}} \sigma^{-2d - |r|}$$

for $r_0 + |r| \leq 2$. Again the resulting bounds are significantly better than the ones stated in Theorem 3.3.3. This finishes the proof for the case where $\alpha = 0$.

As for the case $d = 1$ and *any* α, it requires only a slight adaptation: in the induction formulas for the $\tilde{S}_n^+$'s, there appear expressions $\alpha \partial_\phi \tilde{S}_{n_1} \cdot \partial_\phi \tilde{S}_{n_2}$ which we must bound too. But according to the proof of the lemma above, $(z\omega \cdot \partial_\phi) \circ \tilde{E}^+$ induces a bounded operator on $\mathbb{B}(h)$, whose norm is actually at most κ. Thus we just need to assume $\omega \neq 0$, which features a non-resonance condition (recall that $d = 1$), and to intersect $\Sigma_{u_1, \frac{3\delta}{4}, \Delta\beta}$ with $\{|z| \geq z_1\}$ where z_1 is an arbitrary positive constant. By restricting to this domain all our definitions and estimates, we obtain that $\partial_\phi \circ \tilde{E}^+$ is a bounded operator and

$$\forall n \geq 1, \qquad \|\tilde{S}_n^+\|_h, \ \|\partial_u \tilde{S}_n^+\|_h, \ \|\partial_\phi \tilde{S}_n^+\|_h \leq \mathcal{S}_n,$$

with the same kind of coefficients $\mathcal{S}_n$ as previously. The resulting bounds are again better than the ones announced in the statement of Theorem 3.3.3; we only lose uniformity with respect to z_1—without regret: ultimately, $\varepsilon = z^{-2}$ is meant to be small!

Note that we have presented here a proof which makes use of majorizing series but it can of course be rephrased as an ordinary Picard fixed point method.

3.3.5. How to prove Theorem 3.3.3 in the general case. We do not assume $\alpha = 0$ or $d = 1$ any longer. The idea of this more general proof is to replace the representation of the invariant manifolds as Lagrangian graphs by a parametrization which carries a better control of the dynamics on them. Let us consider again the case of the stable manifold. One can look at and for it as a manifold foliated by the stable manifolds $\mathcal{W}_{\underline{\theta}}^+$ of the individual points $\underline{\theta}$ of $\mathcal{T}_0$, defined by

$$\mathcal{W}_{\underline{\theta}}^+ = \{\, M \in T^*C \mid \mathrm{dist}(\varphi_H^t(M), \varphi_H^t(\underline{\theta})) \xrightarrow[t \to +\infty]{} 0 \text{ exponentially fast}\,\},$$

where we denote by $C =]2\pi - q_0, 2\pi + q_0[\times \mathbb{T}^d$ a part of the configuration space, by T^*C the corresponding part of the phase space, and by φ_H^t the time-t map of the Hamiltonian flow associated to $H_{\varepsilon, \mu}$.

Since $\mathcal{T}_0 = \{\underline{\theta} = (2\pi, \theta, 0, 0), \ \theta \in \mathbb{T}^d\}$, $\varphi_H^t((2\pi, \theta, 0, 0)) = (2\pi, \theta + zt\omega, 0, 0)$ (with $z = \varepsilon^{-1/2}$) and the unperturbed situation is clear, looking for the stable foliation amounts to looking for functions P, Φ_j, J_j of (u, ϕ, z, μ), with exponential decrease when u goes to $+\infty$, such that the manifold parametrized by

(6) $$\begin{cases} q = q_0(u) \\ p = \dot{q}_0(u) + \mu P(u, \theta; z, \mu) \\ \phi = \theta + \mu \Phi(u, \theta; z, \mu) \\ I = \mu J(u, \theta; z, \mu) \end{cases}$$

is invariant under the flow and the pull-back of the Hamiltonian vector field restricted to it can be written:

$$\begin{cases} \dot{u} = 1 + O(\mu) \\ \dot{\theta} = z\omega. \end{cases}$$

We will not write the equations one obtains for P, Φ, J, but only mention that they involve the operators $\tilde{D}_0$ and ∂_u but *not* partial differentiation with respect to the angles ϕ_j: the difficulty that we alluded to at the beginning of this paragraph does not show up, and we can find real analytic solutions P, Φ, J, e.g. by using the ordinary Banach fixed point theorem. Our method is in fact a close cousin of the usual one for proving the stable manifold theorem in finite differentiability (see [Y2], [HPS] and above Chapter 1, §1.2). It is in fact even somewhat easier in our case since transverse regularity, *i.e.* the analyticity with respect to θ, comes for free. To be completely specific, one can determine μ_2 such that these functions are analytic for

$$u \in \mathcal{D}^+_{u_1,\delta},\ \theta \in \mathbb{T}^d_{h_0-\frac{\sigma}{2}},\ z \in \Sigma_{u_1,\delta,\Delta\beta},\ |\mu| < \mu_2.$$

Then the inversion problem:

$$\phi = \theta + \mu\Phi(u,\theta;z,\mu) \quad \Leftrightarrow \quad \theta = \phi + \mu\Theta(u,\phi;z,\mu)$$

can be solved in order to eliminate θ from the parametrization (6) of $\mathcal{W}^+$:

$$\begin{cases} q = q_0(u) \\ p = \dot{q}_0(u) + \mu P(u,\phi+\mu\Theta(u,\phi;z,\mu);z,\mu) = \mathcal{P}(u,\phi;z,\mu) \\ \phi = \phi \\ I = \mu J(u,\phi+\mu\Theta(u,\phi;z,\mu);z,\mu) = \mathcal{J}(u,\phi;z,\mu), \end{cases}$$

and there remains only to perform an integration, for μ small enough, in order to recover the function $\tilde{S}^+$. We refer again the interested reader to [Sa2] (§5) where the proof is given at great length.

3.4. Exponential closeness of the invariant manifolds

Our goal now is to estimate the difference of the generating functions:

$$\Delta S = \sum_{n\geq 1} \mu^n \Delta S_n, \quad \text{where} \quad \Delta S_n = S_n^+ - S_n^- \quad \text{for } n \geq 1.$$

When considering the Fourier expansion of a function on $\mathbb{T}^d$, we denote as usual the Fourier modes by an index k running over $\mathbb{Z}^d$; for instance:

$$F(q,\phi) = \sum_{k\in\mathbb{Z}^d} F_k(q)\, e^{ik\cdot\phi}.$$

Following the notation of §3.3.2 we get as a starting point the following bounds, valid for all $\delta, \sigma > 0$ with $\delta < \pi/2$ and $\sigma < h_0$ and all $u \in \mathcal{C}_\delta$:

$$|\tilde{F}_k(u)| \leq A(\delta,\sigma)\, e^{-2|\Re e\, u|}\, e^{-(h_0-\sigma)|k|}.$$

We recall that the tilde indicates the substitution $q = q_0(u)$.

3.4.1. Using (3) it is easy to compute and estimate the Fourier coefficients of the Poincaré-Melnikov approximation (compare [L1] or [L2], p. 118). Explicitly we have, for any $k \in \mathbb{Z}^d$:

(7) $$\Delta\tilde{S}_{1,k}(u;\varepsilon) = \int_{-\infty}^{+\infty} \tilde{F}_k(u+\zeta)\, e^{i\varepsilon^{-1/2}\zeta\omega\cdot k}\, d\zeta.$$

We observe that the mean value is independent of u:

$$\Delta\tilde{S}_{1,0}(u;\varepsilon) = a_1(\varepsilon).$$

The set $\mathcal{C}_\delta$ contains the horizontal strip of width $\pi - 2\delta$ centered on the real axis. Therefore in the integral (7) we can push the path of integration upwards or downwards up to a distance $\rho = \frac{\pi}{2} - \delta$ from the real axis when u is real. So for any $\varepsilon > 0$

and any $u \in \mathbb{R}$ we have:
$$\Delta \tilde{S}_{1,k}(u;\varepsilon) = -e^{-\rho \varepsilon^{-1/2}(\pm \omega \cdot k)} \int_{-\infty}^{+\infty} \tilde{F}_k(u \pm i\rho + \xi) \, e^{i\varepsilon^{-1/2}\xi \omega \cdot k} \, d\xi,$$

and by selecting the sign $\pm$ according to the sign of $\omega \cdot k$ we immediately obtain the following estimates:

PROPOSITION. *For any $\varepsilon > 0$, $u \in \mathbb{R}$ and $k \in \mathbb{Z}^d$, the k-th Fourier coefficient of the Poincaré-Melnikov integral satisfies the inequality:*
$$|\Delta \tilde{S}_{1,k}(u;\varepsilon)| \leq A(\delta, \sigma) \, e^{-\rho \varepsilon^{-1/2}|\omega \cdot k| - h|k|}, \quad \text{with} \quad \rho = \frac{\pi}{2} - \delta, \quad h = h_0 - \sigma.$$

What about the function $\Delta \tilde{S}_1$ itself? We need to bound sums of exponentials, which requires a bit of work since the potentially small "divisors" $|\omega \cdot k|$ are involved. It is actually quite easy to get an *upper bound*, as embodied in the lemma below. It is however much more difficult to study the possible optimality of this upper bound, a problem to which we return in Paragraph 3.5. As for now we just state an easy lemma about sums of exponentials; note that this is independent of the perturbation F.

LEMMA. *Suppose that $\omega \in \mathbb{R}^d$ satisfies the Diophantine condition*
$$\forall k \in \mathbb{Z}^d \setminus \{0\}, \quad |\omega \cdot k| \geq \gamma |k|^{1-\tau}$$

for some fixed numbers $\gamma > 0$ and $\tau \geq d$. Let $\nu = (1 + (\tau - 1)^{-1})((\tau - 1)\gamma)^{1/\tau}$ and $w(\delta, \sigma) = \nu \left(\frac{\pi}{2} - \delta\right)^{\frac{1}{\tau}} (h_0 - 2\sigma)^{\frac{\tau-1}{\tau}}$, for $0 < \delta < \frac{\pi}{2}$ and $0 < \sigma < h_0/2$.

With these assumptions and notation there exists a positive constant $c = c(d)$ depending only on the dimension such that:

$$(8) \qquad \sum_{k \in \mathbb{Z}^d \setminus \{0\}} e^{-\rho \varepsilon^{-1/2}|\omega \cdot k| - h|k|} \leq \frac{c}{\sigma^d} \exp(-w(\delta, \sigma) \, \varepsilon^{-\frac{1}{2\tau}}),$$

where $\rho = \frac{\pi}{2} - \delta$ and $h = h_0 - \sigma$.

This lemma enables us to deduce from the previous proposition exponentially small bounds for $\Delta \tilde{S}_1$ in the real domain, as well as its partial derivatives, still under the assumption that ω satisfies a Diophantine condition. The smaller δ and σ, the larger the width $w(\delta, \sigma)$. One can even reach $w_* = w(0, 0)$ by taking $\delta = \sigma = \varepsilon^{\frac{1}{2\tau}}$: indeed with that choice we find that $w(\delta, \sigma) \varepsilon^{-\frac{1}{2\tau}} \geq w_* \varepsilon^{-\frac{1}{2\tau}} - \text{cst}$, which leads to the following

COROLLARY. *Suppose that $\omega \in \mathbb{R}^d$ satisfies the Diophantine condition of the preceding lemma and let*
$$w_* = (1 + (\tau - 1)^{-1})((\tau - 1)\gamma)^{\frac{1}{\tau}} \left(\frac{\pi}{2}\right)^{\frac{1}{\tau}} h_0^{\frac{\tau-1}{\tau}}.$$

Then there exist positive constants ε_0 and b such that, for all $(r_0, r) \in \mathbb{Z}_+ \times \mathbb{Z}_+^d$ with $r_0 + |r| \leq 2$,

$$(9) \quad |(\partial_u)^{r_0}(\partial_\phi)^r (\Delta \tilde{S}_1(u, \phi; \varepsilon) - a_1(\varepsilon))| \leq b \, A(\varepsilon^{\frac{1}{2\tau}}, \varepsilon^{\frac{1}{2\tau}}) \varepsilon^{-\frac{r_0 + |r| + d}{2\tau}} \exp(-w_* \varepsilon^{-\frac{1}{2\tau}}),$$

whenever $(u, \phi) \in \mathbb{R} \times \mathbb{T}^d$ and $0 < \varepsilon \leq \varepsilon_0$ (recall that $a_1(\varepsilon)$ denotes the u-independent mean value of $\Delta \tilde{S}_1$).

Observe that the analyticity width h_0 enters this upper bound precisely through the width w_* (hence the name). Observe also that if F is a trigonometric polynomial, *i.e.* if $F_k = 0$ for $|k|$ large enough, *no small divisors occur* and one obtains a much *smaller* upper bound, with exponent $1/2$ *independent of τ and the dimension d* instead of $1/(2\tau)$. This is precisely the phenomenon which was noted in [L1] and the reason why Hamiltonian ($*$) was introduced there (in a slightly less general version; *cf.* §3.1.2 *in fine*). We return to these questions in the next paragraph.

REMARK. As an aside we note that in the present paper we prefer to write the standard polynomial Diophantine condition as above, resulting in a shift of the exponent by 1 as compared to many papers (including [L1], [L2]). The formulas come out neater with the present normalization and this is no chance phenomenon. It simply has to do with the discrepancy between homogeneous and inhomogeneous approximations, or if one wants with the fact that we could rescale one of the nonzero coordinates of ω to 1, resulting in a shift in the effective dimension. Basically for the same reason the present normalization is also the correct one for the statement of transfer properties between linear and simultaneous approximations (see [L2], Appendix 1).

3.4.2. A lemma on Fourier coefficients.

Such direct arguments can also be given in order to bound $\Delta \tilde{S}_2$ using Formula (4) and one could try and build a "direct" machinery in order to bound all the individual $\Delta \tilde{S}_n$, but there is a much better and more efficient way, stemming from the work of V.F.Lazutkin in the one-frequency case. We will return to some historical remarks on this in §3.7.2. In fact the results on exponential smallness that we will obtain in the sequel will essentially follow from the combination of the analyticity Theorem 3.3.3 with the following lemma on the unperturbed characteristic vector field D_0, or rather its straightened version $\tilde{D}_0$ (*cf.* §§3.2.3–3.2.4).

LEMMA. *Let $\varepsilon > 0$. Suppose a function $\tilde{\chi}(u, \phi)$ is analytic in $]-i\rho_0, i\rho_0[\times \mathbb{T}^d_{h_0}$ for some $\rho_0, h_0 > 0$ and satisfies the linear homogeneous partial differential equation $\tilde{D}_0 \tilde{\chi} = 0$. Then $\tilde{\chi}$ extends analytically to $\{|\Im m \, u| < \rho_0\} \times \mathbb{T}^d_{h_0}$ and its Fourier coefficients with respect to the angles ϕ_j satisfy the following inequalities, for all positive $\rho < \rho_0$ and $h < h_0$, all $k \in \mathbb{Z}^d$ and all $u \in \mathbb{R}$:*

$$|\tilde{\chi}_k(u)| \leq \left(\sup_{[-i\rho, i\rho] \times \overline{\mathbb{T}}^d_h} |\tilde{\chi}| \right) e^{-\rho \varepsilon^{-1/2} |k \cdot \omega| - h|k|}.$$

This is the quasiperiodic generalization of the one-frequency version in [La1]. It appears basically in this form in [DGJS]. The extension to the quasiperiodic case is actually essentially trivial because this is actually a lemma in the theory of complex functions in *one* variable. Its first part is an extension statement for the solutions of a particularly simple PDE, and the second one gives a control on the coefficients. Mathematically speaking this is all fairly elementary and amounts in effect to little more than shifting contours of integration. We insist on this because the *real* idea, due to V.F.Lazutkin, consists in making use of this lemma under the present cirumstances in order to exploit the following principle: if a function of one complex variable is analytic over a (large) open strip centered on the real axis and can be extended to a bounded function on the closure, one has a good control of the function *on* the real axis.

The lemma applies to the Poincaré-Melnikov integral $\Delta \tilde{S}_1$ itself, since $D_0 \Delta S_1$ vanishes, or equivalently $\tilde{D}_0 \Delta \tilde{S}_1 = 0$, and one can recover Proposition 3.4.1 in this way. Using a strategy which seems to be a novel feature of the present paper, we will now see how to use the above lemma in order to bound the *whole* function ΔS, at the price of only a change of coordinates. This change of coordinates f is necessary since there is no reason why D_0 itself should annihilate the function ΔS: this will however be the case for the function $\Delta \Sigma = \Delta S \circ f$.

In fact, anticipating a little, we first geometrically define in §3.4.3 a vector field D which annihilates ΔS and turns out to be the direct image of D_0 by some diffeomorphism f; this is the technical work subtending §3.4.4. Then in §§3.4.5–3.4.6 we state two theorems which are not difficult, given all the preparatory work: the first one yields exponentially small bounds for the Fourier coefficients of $\Delta \Sigma$ which are obtained by the above lemma; the second one combines this same lemma with the upper bound given by Lemma 3.4.1 (it thus requires a Diophantine condition on ω) to yield exponentially small bounds for the partial derivatives of ΔS.

3.4.3. Characteristic vector fields. Since the geometric tool that we introduce here is not specific to our model Hamiltonian (∗) we will temporarily widen our framework, considering a differentiable manifold M (the configuration space) and a function $H : T^*M \to \mathbb{R}$ on its cotangent bundle (the phase space). This cotangent bundle is endowed as usual with its canonical exact symplectic structure, induced by the Liouville one-form λ. We denote by $\pi : T^*M \to M$ the natural projection and by X_H the Hamiltonian vector field generated by H. What we do below is actually too general for our purpose and all we need is to construct D and show that it annihilates ΔS. This is the content of Proposition 1.2 in [Sa2], whose proof is quite short. Yet we feel that dwelling a bit on the geometry may be enlightening and certainly makes the construction look more natural.

If α is a one-form on M, we denote by $\text{im}(\alpha)$ or $\mathcal{G}r(\alpha)$ its image on T^*M, viewing α as a section of the projection π. It is a submanifold of T^*M and π induces a diffeomorphism between $\text{im}(\alpha)$ and M, which identifies α with the restriction of λ to $\text{im}(\alpha)$. This last property actually characterizes the Liouville form λ. Moreover $\text{im}(\alpha)$ is Lagrangian if and only if α is closed, and is exact Lagrangian if and only if α is exact (see also §§1.1.13 and 1.7.1).

If α is closed, $\text{im}(\alpha)$ is invariant by the Hamiltonian vector field X_H if and only if H is constant on it, *i.e.* if and only if the function $H \circ \alpha$ is constant on M: this is the Hamilton-Jacobi equation. In that situation the *characteristic vector field* of $\text{im}(\alpha)$ is usually defined to be the vector field on M which is the direct image by π of the restriction of the Hamiltonian vector field to $\text{im}(\alpha)$. It may be written $T\pi \circ X_H \circ \alpha$ (*cf.* §1.1.15). We propose the following generalization of that construction:

DEFINITION. Given any pair (α_0, α_1) of one-forms of M, we call characteristic vector field of (α_0, α_1) the vector field on M obtained as

$$D = \int_0^1 D_t \, dt \quad \text{where, for } 0 \leq t \leq 1, \ D_t = T\pi \circ X_H \circ \big(\alpha_0 + t(\alpha_1 - \alpha_0)\big).$$

In the exact case, if $\alpha_0 = dS_0$, $\alpha_1 = dS_1$, with S_0, S_1 functions on the configuration space, D will also be called the characteristic vector field of the pair (S_0, S_1).

In a local coordinate system $(Q_1, \ldots, Q_n)$ of M we can write:

$$D = \sum_{1 \leq j \leq n} D_j \frac{\partial}{\partial Q_j}, \qquad D_j(Q) = \int_0^1 \frac{\partial H}{\partial P_j}(\alpha_t(Q)) \, dt,$$

setting $\alpha_t = (1-t)\alpha_0 + t\alpha_1$ for $0 \leq t \leq 1$ and using the induced canonical system of coordinates (Q, P) on T^*M.

The interest of this construction lies in the following proposition (the easy proof of which we omit), which shows that D somehow reflects in the configuration space what happens in the phase space:

PROPOSITION. *Let D be the characteristic vector field of a pair (α_0, α_1) of one-forms on M. Then its action on the difference $\alpha_1 - \alpha_0$ can be described as:*

$$< \alpha_1 - \alpha_0, D > = H \circ \alpha_1 - H \circ \alpha_0.$$

If in particular α_0 and α_1 satisfy the Hamilton-Jacobi equation associated with the same energy level, this quantity vanishes on M.

If the Hamiltonian function is quadratic in the momenta $P_1, \ldots, P_n$, the vector field D is merely the arithmetic mean of D_0 and D_1, the ordinary characteristic vector fields associated to S_0 and S_1. In the case of the Hamiltonian $(*)$ and of the generating function $S^\pm$ we thus simply obtain:

$$D = \frac{1}{2}\partial_q(S^+ + S^-)\frac{\partial}{\partial q} + (\varepsilon^{-1/2}\omega + \frac{1}{2}\alpha\partial_\phi(S^+ + S^-)) \cdot \frac{\partial}{\partial \phi}.$$

So at this point we did use in a significant way the fact that the Hamiltonian is quadratic in the action, which for Hamiltonian $(*)$ is the same as requiring that the perturbation be independent of the action variables, which in turn is part of Assumption 3.1.2 i). So under our current assumptions we find by the proposition above that $D\Delta S = < d(\Delta S), D > = 0$.

As was sketched above this fact has important consequences here, since our goal is to study the function ΔS defined on part of the configuration space $M_1 = \,]2\pi - q_1, q_1[\times \mathbb{T}^d$ for $\varepsilon > 0$ and μ small enough (q_1 is arbitrary and fixed such that $\pi < q_1 < 2\pi$). We now discover that this function is constant along the integral curves of the vector field D. Especially we notice that $D = D_0 + O(\mu)$ and thus any torus $\{q_*\} \times \mathbb{T}^d$ inside M_1 is transverse to D for μ small enough, as it is certainly transverse to D_0. The function ΔS is determined by its restriction to any given such torus and the critical points of the restriction yield critical points of ΔS itself. Any function on the torus $\mathbb{T}^d$ has at least $d+1$ critical points by the Ljusternik-Schnirelman theorem. Hence we easily deduce the following

COROLLARY. *There exists $\mu_0' > 0$ such that for $\varepsilon > 0$ and $|\mu| \leq \mu_0'$, the Hamiltonian system associated to $H_{\varepsilon,\mu}$ admits at least $d+1$ distinct homoclinic orbits.*

This is not a new result by any means and it was stated rather to show how some notions transcribe in this setting; see in fact §1.10 for a more general discussion and references. As is the case for any flow and as was already discussed in Chapters 1 and 2, the Hessian matrix of ΔS at a critical point (q_*, ϕ_*) is always degenerate and it is rather the Hessian matrix of the restriction $\Delta S_{|q=q_*}$ which provides a symplectic measure of the splitting along the corresponding homoclinic orbit.

3.4.4. Straightening the characteristic vector field.
As explained earlier, in order to use Lemma 3.4.2 together with the property $D\Delta S = 0$ we need to conjugate D to D_0. This amounts to straightening D, *i.e.* to finding coordinates in which it has constant coefficients. The existence of such a kind of global flow-box coordinates in the real domain is not surprising, since we already noticed the existence of global sections of the configuration space which are transverse to D. But we need to define the change of coordinates in a complex domain, which moreover should be as large as possible. Again it will be convenient to make use of the variable u rather than q. This is technical work which we only summarize here, referring as usual to [Sa2] for details.

Let $u_2 > 0$, $\Delta\beta \in]0, \arctan(\pi/2u_2)[$ and $\delta, \sigma > 0$ as in Theorem 3.3.3. The variable u runs over $\mathcal{D}_{u_2,\delta} = \mathcal{D}^+_{u_2,\delta} \cap \mathcal{D}^-_{u_2,\delta}$ and the variable $z = \varepsilon^{-1/2}$ over $\overline{\Sigma}_{u_2,\delta,\Delta\beta}$. Observe that $\mathcal{D}_{u_2,\delta}$ is a lozenge with corners u_2, $i\rho'$, $-u_2$ and $-i\rho'$, where $\rho' = \pi/2 - \delta/\cos\beta_2$ and we denote by β_2 the half-aperture of the sectors $\mathcal{D}^-_{u_2,\delta}$ and $\mathcal{D}^+_{u_2,\delta}$; see Figure 1 in §3.3.3 and recall that $\mathcal{D}^-_{u_2,\delta}$ and $\mathcal{D}^+_{u_2,\delta}$ are opposite sectors. In particular, we will be able to ensure that

$$[-i\rho, i\rho] \subset \mathcal{D}_{u_2,\delta}, \qquad \rho = \frac{\pi}{2} - 2\delta,$$

simply by taking $u_2 \geq \frac{\pi}{2\sqrt{3}}$. Then we have:

PROPOSITION. *There exist a positive number μ_2 and a real analytic change of coordinates*

$$(u, \phi) = (v, \theta) + \mu\mathcal{U}(v, \theta; z, \mu) \quad \Leftrightarrow \quad (v, \theta) = (u, \phi) + \mu\mathcal{V}(u, \phi; z, \mu)$$

satisfying the following properties:

a) $\mathrm{Id} + \mu\mathcal{U}$ *conjugates the characteristic vector field $\tilde{D}$ with $\frac{\partial}{\partial v} + z\omega \cdot \frac{\partial}{\partial \theta}$;*

b) $\mathrm{Id} + \mu\mathcal{U}$ *induces a bijection between the domain $\overline{\mathcal{D}}_{u_2,\delta} \times \overline{\mathbb{T}}^d_{h_0-\sigma}$ and its image for $|\mu| \leq \mu_2$, $z \in \overline{\Sigma}_{u_2,\delta,\Delta\beta}$; for these values of (v, θ, z, μ) the components of $\mathcal{U}$ are analytic with respect to all their arguments;*

c) $\mathrm{Id} + \mu\mathcal{V}$ *induces a bijection between the domain $\overline{\mathcal{D}}_{u_2,\delta} \times \overline{\mathbb{T}}^d_{h_0-\sigma}$ and its image for $|\mu| \leq \mu_2$, $z \in \overline{\Sigma}_{u_2,\delta,\Delta\beta}$; for these values of (u, ϕ, z, μ) the components of $\mathcal{V}$ are analytic with respect to all their arguments.*

One can even give quantitative information on a possible μ_2 as well as bounds for the partial derivatives of the components of $\mathcal{U}$ and $\mathcal{V}$, in the spirit of Theorem 3.3.3, when $2\delta \leq \sigma$.

Via the change of variable $q = q_0(u)$, the diffeomorphism $\mathrm{Id} + \mu\mathcal{U}$ induces a mapping f according to the formula

$$(u, \phi) = (\mathrm{Id} + \mu\mathcal{U})(v, \theta) \Leftrightarrow (q_0(u), \phi) = f(q_0(v), \theta),$$

which establishes a diffeomorphism between $M_2 =]q_0(-u_2), q_0(u_2)[\times \mathbb{T}^d$ and some domain M_1; this is as far as real domains are concerned, but it extends to some complex domain. This diffeomorphism conjugates D with D_0, so that $D_0.(\Delta S \circ f)$ vanishes as previously announced.

Here are some indications on the proof of Proposition 3.4.4 (see [Sa2] §6 for details): As is the case for the proof of Theorem 3.3.3, it is sufficient to apply the ordinary fixed point theorem in a suitable Banach space. And again as in the proof of Theorem 3.3.3, the keypoint is the existence of a bounded right inverse for $\tilde{D}_0$.

Indeed, the equations to be solved in order to find the components $\mathcal{U}_u, \mathcal{U}_{\phi_1}, \ldots, \mathcal{U}_{\phi_d}$ of $\mathcal{U}$ can be written:

$$\tilde{D}_0 \mathcal{U}_u = \tilde{D}_u \circ (\mathrm{Id} + \mu \mathcal{U}), \qquad \tilde{D}_0 \mathcal{U}_{\phi_j} = \tilde{D}_{\phi_j} \circ (\mathrm{Id} + \mu \mathcal{U}), \qquad j = 1, \ldots, d.$$

Then one checks that for a given convergent Fourier series $\psi = \sum \psi_k(v,z) e^{ik\cdot\theta}$, whose coefficients are analytic in $\mathcal{D}_{u_2,\delta} \times \Sigma_{u_2,\delta,\Delta\beta}$ and continuous on the closure, the formulas

$$\forall k \in \mathbb{Z}^d, \quad \chi_k(v,z) = \begin{cases} -e^{-izvk\cdot\omega} \int_v^{i\rho'} e^{iz\zeta k\cdot\omega} \psi_k(\zeta,z)\,d\zeta & \text{if } k\cdot\omega > 0, \\ e^{-izvk\cdot\omega} \int_{-i\rho'}^{v} e^{iz\zeta k\cdot\omega} \psi_k(\zeta,z)\,d\zeta & \text{if } k\cdot\omega < 0, \\ \int_0^v \psi_k(\zeta,z)\,d\zeta & \text{if } k\cdot\omega = 0, \end{cases}$$

define a function $\chi = \sum \chi_k(v,z) e^{ik\cdot\theta}$ such that $\tilde{D}_0 \chi$ coincides with ψ. The correspondence $\psi \mapsto \chi = E_{\rho'} \psi$ defines a bounded operator $E_{\rho'}$ on an appropriate Banach space in which it is possible to find $\mathcal{U}_u, \mathcal{U}_{\phi_1}, \ldots, \mathcal{U}_{\phi_d}$ such that

$$\mathcal{U}_u = E_{\rho'}\big(\tilde{D}_u \circ (\mathrm{Id} + \mu \mathcal{U})\big), \qquad \mathcal{U}_{\phi_j} = E_{\rho'}\big(\tilde{D}_{\phi_j} \circ (\mathrm{Id} + \mu \mathcal{U})\big), \qquad j = 1, \ldots, d.$$

3.4.5. Exponentially small Fourier coefficients for the difference of the generating functions.
We now state, in this and the next subsection, two theorems which are obtained by simply putting together the previous results. The first one (this subsection) requires no Diophantine condition on ω but it deals with other generating functions than $S^\pm$: a near-to-identity change of coordinates is applied in the configuration space and then lifted to the phase space, so that the invariant manifolds are given by new generating functions $\Sigma^\pm$. The second result (§3.4.6) is stated in terms of the original variables, at the cost of imposing a Diophantine condition and using Lemma 3.4.1 which is not optimal.

THEOREM (Exponentially small Fourier coefficients). *Let $Q_2 \in]\pi, 2\pi[$ and let δ, σ be small enough positive numbers. Let $\rho = \pi/2 - 2\delta$, $h = h_0 - \sigma$ and $M_2 =]2\pi - Q_2, Q_2[\times \mathbb{T}^d$. There exist a subdomain M_1 of $]0, 2\pi[\times \mathbb{T}^d$, an exact symplectic diffeomorphism Φ between T^*M_2 and T^*M_1 and functions Σ^- and Σ^+ on M_2 (which depend analytically on (ε, μ)) such that*

$$\mathcal{W}^\pm = \Phi\big(\mathcal{G}r(d\Sigma^\pm)\big), \qquad \Sigma^+ - \Sigma^- = \mu \Delta S_1 + \chi,$$

where the Fourier coefficients χ_k^∂ of the partial derivatives $\partial \chi$ of the function χ satisfy

$$|\chi_k^\partial(Q; \varepsilon, \mu)| \leq C_\partial |\mu|^2 e^{-\rho \varepsilon^{-1/2}|k\cdot\omega| - h|k|}$$

for $k \in \mathbb{Z}^d$, $Q \in]2\pi - Q_2, Q_2[$, $\varepsilon > 0$, $\mu \in [-\mu_2, \mu_2]$ and $\partial = (\partial_Q)^{r_0} (\partial_\theta)^r$ with $r_0 + |r| \leq 2$. If $2\delta \leq \sigma$, one can take $\mu_2 = b_2^{-1} A(\delta/4, \sigma/4)^{-1} \delta^2 \sigma^d$ and

$$C_\partial = \begin{cases} b_2\, A(\delta/4, \sigma/4)^2 \delta^{-3} \sigma^{-2d-1} & \text{if } r_0 + |r| = 0, \\ b_2\, A(\delta/4, \sigma/4)^2 \delta^{-2-r_0} \sigma^{-2d-1-|r|} & \text{if } r_0 + |r| = 1 \text{ or } 2, \end{cases}$$

where the positive number b_2 depends only on Q_2.

SKETCH OF PROOF. Φ is nothing but the exact symplectic lift of the diffeomorphism $f : M_2 \to M_1$ of §3.4.4 (the corresponding "point transformation", cf. §1.7.3) defined by:
$$\Phi : \begin{cases} T^*M_2 & \to \quad T^*M_1 \\ \beta & \mapsto \Phi(\beta) = {}^t(T_{f\circ\pi(\beta)}f^{-1}).\beta, \end{cases}$$
where π denotes the natural projection $T^*M_2 \to M_2$. Not only Φ is a lift of f (i.e. $\pi \circ \Phi = f \circ \pi$) which preserves the Liouville form λ, but its action on exact Lagrangian graphs is easily described: according to Proposition 1.7.3, if S is a function on M_1,
$$\mathcal{G}r(dS) = \Phi\big(\mathcal{G}r(d\Sigma)\big) \quad \text{with} \quad \Sigma = S \circ f.$$
In particular, $\mathcal{W}^\pm = \Phi\big(\mathcal{G}r(d\Sigma^\pm)\big)$ with $\Sigma^\pm = S^\pm \circ f$. Consider the counterpart $\Delta\tilde{\Sigma}$ in the variable v (with the notation of §3.4.4) of the function $\Delta\Sigma = \Sigma^+ - \Sigma^-$. Since f was chosen precisely in order to yield $D_0 \Delta\Sigma = 0$, i.e. $\tilde{D}_0 \Delta\tilde{\Sigma} = 0$, and since $\tilde{D}_0$ has constant coefficients, we have $\tilde{D}_0 \partial(\Delta\tilde{\Sigma}) = 0$. Recalling that $\tilde{D}_0 \partial(\Delta\tilde{S}_1) = 0$ as well, we can now apply Lemma 3.4.2 to the difference $\partial(\Delta\tilde{\Sigma} - \mu\Delta\tilde{S}_1) = O(\mu^2)$. Moreover this $O(\mu^2)$-remainder can be precisely bounded thanks to Theorem 3.3.3 and the quantitative information we have alluded to after the statement of Proposition 3.4.4. $\square$

3.4.6. We now state the result which is perhaps most immediately relevant for our purpose in this chapter.

THEOREM (Exponential smallness of ΔS). *Suppose that $\omega \in \mathbb{R}^d$ satisfies the usual Diophantine condition*
$$|\omega \cdot k| \geq \gamma |k|^{1-\tau} \quad \text{for all } k \in \mathbb{Z}^d \setminus \{0\},$$
with some fixed numbers $\gamma > 0$ and $\tau \geq d$, and define
$$w_* = \big(1 + (\tau - 1)^{-1}\big)\big((\tau - 1)\gamma\, h_0^{\tau-1}\, \tfrac{\pi}{2}\big)^{\frac{1}{\tau}}.$$
Then for any closed subinterval $[q_1, q_2]$ of $]0, 2\pi[$ there exist positive constants ε_0 and b such that, for $(r_0, r) \in \mathbb{Z}_+ \times \mathbb{Z}_+^d$ with $r_0 + |r| = 1$ or 2,
$$|(\partial_q)^{r_0}(\partial_\phi)^r(\Delta S - \mu \Delta S_1)(q, \phi; \varepsilon, \mu)| \leq$$
$$b|\mu|^2 A(\varepsilon^{\frac{1}{2\tau}}, \varepsilon^{\frac{1}{2\tau}})^2\, \varepsilon^{-\frac{r_0+|r|+3d+3}{2\tau}}\, \exp(-w_* \varepsilon^{-\frac{1}{2\tau}})$$
whenever $(q, \phi) \in [q_1, q_2] \times \mathbb{T}^d$, $0 < \varepsilon \leq \varepsilon_0$, $|\mu| \leq b^{-1} A(\varepsilon^{\frac{1}{2\tau}}, \varepsilon^{\frac{1}{2\tau}})^{-1} \varepsilon^{\frac{d+2}{2\tau}}$.

REMARK. As for the case $(r_0, r) = (0, 0)$, i.e. the difference $\Delta S = S^+ - S^-$ itself, we need to take the average into account and there exists a real analytic function $a(\varepsilon, \mu)$ such that $\Delta S - \mu\Delta S_1 - \mu^2 a$ satisfies the same kind of inequality:
$$|(\Delta S - \mu\Delta S_1)(q, \phi; \varepsilon, \mu) - \mu^2 a(\varepsilon, \mu)| \leq b|\mu|^2 A(\varepsilon^{\frac{1}{2\tau}}, \varepsilon^{\frac{1}{2\tau}})^2\, \varepsilon^{-\frac{3d+4}{2\tau}}\, \exp(-w_* \varepsilon^{-\frac{1}{2\tau}}).$$

SKETCH OF PROOF. Let $\delta, \sigma > 0$ small enough. Apply Lemma 3.4.1 to the function $\partial\chi$ of Theorem 3.4.5 and return to the function $\partial(\Delta S - \mu\Delta S_1)$ using the inverse change of variables f^{-1}. Since this change of variables is real analytic, the bounds one obtains can be transferred from the real part of the (Q, θ)-domain to the real part of the (q, ϕ)-domain for real values of the parameters ε and μ:
$$|\partial(\Delta S - \mu\Delta S_1)| \leq b_3 |\mu|^2 A(\delta/4, \sigma/4)^2 \delta^{-2-r_0} \sigma^{-3d-1-|r|} \exp(-w(\delta, \sigma)\, \varepsilon^{-\frac{1}{2\tau}}).$$

Then one reaches $w_* = w(0,0)$ by taking $2\delta = \sigma = 8\,\varepsilon^{\frac{1}{2\tau}}$. This completes this very sketchy sketch of proof (see as usual [Sa2] for details). □

Comparing this theorem and Corollary 3.4.1, we see that we have obtained upper bounds for the distance between the invariant manifolds, with exponent $1/(2\tau)$ and width w_*. We also get bounds on the Hessian matrix of the $\Delta S_{|q=q_*}$ at a homoclinic point. All this was already derived in Chapter 2, in a more general case and a more geometric environment. The novelty here consists of course in the estimate of the difference between the full and the linearized quantities, whether it be for the distance between the invariant manifolds $\mathcal{W}^\pm$ or the splitting matrix at homoclinic points (or even for the difference of the generating functions—see the remark above). Now there remains of course the problem of estimating the *linear* part, namely ΔS_1 and its derivatives. All we have at our disposal at this point is Lemma 3.4.1 which is an easy upper bound but does not go deep into the arithmetic. One of the messages in [L1] was indeed that the evaluation of the linear part of the splitting is a difficult problem. In other words the question of the possible optimality of Lemma 3.4.1, which first has to be properly defined and worded (see §3.5 below), seems to be quite difficult to attack in general. Yet it appears as a prerequisite if one wishes to derive lower bounds (again this has to be carefully phrased; see §3.5) for the distance between and the splitting of the invariant manifolds, with of course μ *polynomially* small in ε.

Rephrasing the above slightly, and somewhat paradoxically perhaps at first sight, it appears that, given the results obtained above, the main problem for deriving *lower* bounds now lies in estimating the *linear* part ΔS_1 (and its derivatives, but this is no serious difference). If one could give a general argument for obtaining a better bound on ΔS_1, it would now be easy to incorporate it into the method in order to bound the remainder $\Delta S - \mu \Delta S_1$, and this would lead to a range of values of μ where the Poincaré-Melnikov approximation does dominate. We will derive results of this kind in the next paragraph, but they are very restrictive and in particular specific to the case of three degrees of freedom ($\ell = 3$).

Lastly we mention that we did not pay much attention to the prefactors (the quantities in front of the exponentials) in the upper bounds derived above: they could have been slightly decreased. But we have always in mind the example of a function $A(.,.,.)$ such that $A(\varepsilon^{\frac{1}{2\tau}}, \varepsilon^{\frac{1}{2\tau}})$ is some power of ε, and we are precisely interested in estimates up to some power of ε. Clearly there is a strong hierarchy in accuracy from the exponent (the roughest of all) to the width and then the prefactor. Also there is an inverse hierarchy in the universality of these quantities: the exponents exhibit strong universality (or genericity if one wants) properties, whereas the widths and *a fortiori* the prefactors are much more sensitive to individual features of the problem. Yet we note that the function A does describe *some* universality, inasmuch as it reduces the information to the rough type of the singularities of the complex extension of the perturbation. In any case we are still very much working just at the level of exponents in these multifrequency problems, as will be illustrated in the next section. Perhaps the reader should recall the case of the one-dimensional WKB method, in which the problem of the prefactors (embodied *e.g.* in reflection coefficients) is quite tricky and not completely understood although one is dealing with a *one*-degree-of-freedom *linear* problem.

3.5. Linear versus nonlinear splitting

As mentioned already at the beginning of §2.6.6, one basic intuition in the domain can now roughly be phrased by saying that the four exponents of the four exponentially small quantities are expected to coincide. This is somehow already visible in [A1] and is developed at a heuristic level in [C] (see also [CV1,2]); we urge the reader to try and penetrate §7.4 of [C], which is a real masterpiece of mathematical physics. We are not going to discuss here the connection between the size of the splitting and the speed at which instability ("Arnold diffusion") develops, referring to [C], [L2] (§V.2) and [L4] for indications, as well as to future works. Instead we concentrate on the comparison between the size of the splitting and that of the linearized splitting ("Poincaré-Melnikov"). We will proceed from the general to the particular, first summarizing some general reasonings and open problems. We then come to the little which is actually known to-date and use it in conjunction with the general results above to derive lower bounds for the splitting in very particular cases.

3.5.1. We start with a problem in Fourier series which hopefully captures a sizable part of the difficulties but can be stated without any reference to the splitting problems.

PROBLEM. Let $\alpha = (\alpha_k)$ be a sequence of real numbers indexed by $k \in \mathbb{Z}^d \setminus \{0\}$, $d \geq 2$; assume that $0 < c < |\alpha_k| < C$ with two fixed constants c and C. Let $\sigma > 0$ be a number and $\varepsilon > 0$ a parameter. Let $\omega \in \mathbb{R}^d$ be a Diophantine vector, satisfying the usual polynomial condition, *i.e.* such that $|\omega \cdot k| > \gamma |k|^{1-\tau}$ for some constants $\gamma > 0$, $\tau \geq d$.

Define the (absolutely convergent) series $S_\alpha(\varepsilon)$ as:

$$S_\alpha(\varepsilon) = \sum_{k \in \mathbb{Z}^d \setminus \{0\}} \alpha_k \exp(-\sigma |k| - \frac{|\omega \cdot k|}{\sqrt{\varepsilon}}).$$

The problem is to study $S_\alpha(\varepsilon)$ as $\varepsilon \to 0$, possibly proving that it is exponentially small with exponent $a = 1/(2\tau)$.

Here we use $\sqrt{\varepsilon}$, rather than just ε, as a small parameter simply because it is the quantity which occurs in the dynamical system problem; the reader may replace $\sqrt{\varepsilon}$ with ε if he finds that it looks better, getting a conjectural exponent $1/\tau$. Note that the fact that $\sqrt{\varepsilon}$ rather than ε is the natural dynamical parameter stems from the fact that Newton's fundamental law ($f = ma$) is a second-order relation. The setting of the problem corresponds to a perturbation F with exact analyticity width $\sigma > 0$ and the α_k's are thought of as $\alpha_k = \exp(\sigma|k|)F_k$, where the F_k's are the Fourier coefficients of F, which is assumed to have zero average ($F_0 = 0$). Things are a bit more complicated in the actual splitting problem, but this seems to capture the essence of the difficulty.

The fact that $S_\alpha(\varepsilon)$ is exponentially small with exponent $\geq 1/(2\tau)$ is not difficult and this is in fact the main content of Lemma 3.4.1 above. The problem consists in showing that the exponent is not strictly bigger; in other words, referring to the introduction of Chapter 2, it can be rephrased as the conjectural assertion that for any $a > 1/(2\tau)$:

$$\limsup_{\varepsilon \to 0} \exp(\varepsilon^{-a}) S_\alpha(\varepsilon) = \infty.$$

The formal computations in [L1] (reproduced in [L2], pp. 120–121) are meant to make this exponent "plausible". Note that if one assumes that the F_k's are positive, compensations cannot occur anymore and things become significantly simpler, although in general quite unrealistic for applications. This remark is however beautifully exploited in [Be2] (Lemma 1.2). Also one does *not* expect that there will be any kind of nice prefactor, nor even actually a nice constant width. In other words, the function $S_\alpha(\varepsilon)$ will "look like" $\exp(w(\varepsilon)/\varepsilon^a)$, where $w(\varepsilon)$ will be quite complicated; yet the decomposition of the argument of the exponential in terms of exponent and width, that is as $w(\varepsilon)/\varepsilon^a$ is still meaningful, as mentioned in the introduction to Chapter 2.

Here we have assumed the best possible situation, *i.e.* that the sequence (α_k) is bounded in norm from above and below, which translates in particular into a no gap condition on the Fourier series of the perturbation F for the Hamiltonian problem. At the other end of the large spectrum of possibilities, if F is a polynomial, then the α_k are almost all zero, the sum $S_\alpha(\varepsilon)$ is *finite* and the exponent is just $1/2$ whatever the value of the Diophantine exponent τ, certainly contradicting the conjectural value. If the statement holds true, it is likely that much less stringent conditions on the sequence (α_k) suffice to ensure its validity but we have not tried to dig any deeper in that direction. Condition ii) in Theorem 3.5.4 represents a very simple example of this phenomenon.

In fact *the problem above is wide open for $d > 2$*, which is a bit worrying as it is really a minimal and very rough preliminary statement in the study of the linearized splitting. The difference between the cases $d = 2$ and $d > 2$ consists as usual in the fact that only in the former case can one make use of the efficient apparatus of (ordinary) continued fractions. Soon enough we will have to confine to that case, which is at present the only one where anything is known for sure. But we will presently dwell a bit on the general case, with some heuristic remarks.

The main remark is that again we are running into apparently big trouble when using *linear* approximation. The whole theory of the splitting at present relies heavily on Fourier series, going along with "small divisors", *alias* linear approximation. One is tempted to say that there must be another way, more in the line of periodic orbits and simultaneous approximation; we have seen some clues for that already, and some more will come, although this is all still very conjectural. But even restricting to the above problem, which is apparently completely concerned with linear approximation, we note that when $d > 2$ one may predict on a heuristic basis that the best Dirichlet approximations of ω and their periods will perhaps play an important role, much as in the "intermittency" phenomenon proposed in [L2] (see especially [L3], §4). The point is to predict a critical sequence (ε_i) of values of the parameter (tending to 0) at which "something happens", namely a shift in the dominant mode(s) of the series $S_\alpha(\varepsilon)$. In the case $d = 2$, where there is no difference between linear and simultaneous approximation, this question has been studied on examples in [S], with a very good agreement between the theory and numerical experiment (but no rigorous results). Simultaneous approximation also strongly suggests which cases would be next in line if one wishes to undertake a case study for $d = 3$. We refer to [L2] pp. 107–113 for an arithmetical discussion which is relevant in the case of the splitting problem.

3.5.2. We now confine ourselves to the case $d = 2$ ($\ell = 3$) for the rest of this paragraph and start with the arithmetic in that case. Without loss of generality one can rescale the first component of the two-dimensional Diophantine vector ω to unity (*e.g.* by rescaling ε) and write $\omega = (1, \chi)$ for a τ-Diophantine *scalar* χ ($\tau \geq 2$). In this situation the only general study we know of is contained in [RW1] §6.3. That section is independent and in particular it is not affected by the errors which have been found in other parts of the paper; it is reproduced and detailed in [RW2]. The authors point out in their §6.3.1 that it is in principle *not* restricted to badly approximable numbers, which correspond to $\tau = 2$ (and are also called numbers of constant type because they are characterized by the fact that the digits of their continued fractions are uniformly bounded). In principle the results reported below (§§3.5.3–3.5.5) *should* extend to the cases examined in [RW1,2] by applying the results obtained there and *should* then match with Theorem 2.3 of [RW1], but we have not checked these assertions.

So instead we narrow the framework even more in order to use the arithmetical analysis in [DGJS]. *In the rest of this paragraph we assume that $\tau = d = 2$ and indeed that $\omega = (1, \chi)$, where χ is the golden number, $\chi = \frac{1+\sqrt{5}}{2}$*. Before being completely specific, we still mention that the paper [DGJS] is devoted to a Hamiltonian of type (∗) with (in our notation) $d = 2$, $\omega = (1, \chi)$, $\alpha = 0$ (the isochronous case), $F = (\cos q - 1)f(\phi)$. The arithmetical study in §2 there clearly displays the features one hopes for in more general cases, namely the fact that only a small number of modes dominate the Poincaré-Melnikov for a given (small) value of ε. This was one of the main ideas underlying the heuristic reasoning in [C] and the formal computation in [L1,2]. It can also be found in [S] in a more precise version, but restricted to the two-dimensional case and still not rigorous. In turn the "transition lemma" in [RW1] is again an avatar of this phenomenon but indeed it gives it an exact rigorous expression for a much larger class of numbers. We mention that the results below could probably be easily extended (without using [RW1,2]) to the (real) quadratic irrationals, *i.e.* the numbers whose continued fraction is eventually periodic. This should in principle be a mild (and not terribly exciting) extension from the golden number itself, but we have not worked it out in details.

We start with a lemma which improves on the result of Lemma 3.4.1 in this very particular case. In order to state it we first introduce some

NOTATION. We denote by $\{x\}$ the distance from a real number x to $\mathbb{Z}$ (so that $0 \leq \{x\} \leq 1/2$). For any $\rho, h > 0$ we denote by $w_{\rho,h}$ the continuous $(4 \log \chi)$-periodic function of the real variable x defined by the formula

$$w_{\rho,h}(x) = C_{\rho,h} \cosh\left(\left\{\frac{x - x_{\rho,h}}{4 \log \chi}\right\} \log \chi\right),$$

where $C_{\rho,h} = 5^{-1/4} \chi \sqrt{4\rho h}$ and $x_{\rho,h} = 2 \log\left(\frac{\rho\sqrt{5}}{h}\right)$.

LEMMA. *There exists $c > 0$ such that, for all $\rho \in]\pi/4, \pi/2[$ and $h \in]h_0/2, h_0[$,*

$$\sum_{k \in \mathbb{Z}^2 \setminus \{0\}} e^{-\rho \varepsilon^{-1/2}|\omega \cdot k| - h|k|} \leq c \exp(-w_{\rho,h}(\log \varepsilon) \varepsilon^{-1/4}).$$

For the proof, which requires only some knowledge of the rational approximations of the golden mean χ, that is some elementary properties of the Fibonacci numbers, the reader is referred to [DGJS] (§§6–7) or [Sa2] (§4.2).

3.5.3. Using this lemma instead of Lemma 3.4.1 in the same chain of reasonings as above, we obtain the following improvement of Corollary 3.4.1 and Theorem 3.4.6:

THEOREM. *Let us use the notation* $w_* = w_{\pi/2, h_0}$. *For any closed subinterval* $[q_1, q_2]$ *of* $]0, 2\pi[$ *there exist positive constants* ε_0 *and* b *such that, for* $(r_0, r) \in \mathbb{Z}_+ \times \mathbb{Z}_+^2$ *with* $r_0 + |r| = 1$ *or* 2,

$$|(\partial_q)^{r_0}(\partial_\phi)^r \Delta S_1| \leq b\, A(\varepsilon^{1/4}, \varepsilon^{1/4})\, \varepsilon^{-\frac{r_0+|r|}{4}} \exp(-w_*(\log \varepsilon)\, \varepsilon^{-1/4})$$

and

$$|(\partial_q)^{r_0}(\partial_\phi)^r (\Delta S - \mu \Delta S_1)(q, \phi; \varepsilon, \mu)| \leq$$
$$b|\mu|^2 A(\varepsilon^{1/4}, \varepsilon^{1/4})^2\, \varepsilon^{-\frac{r_0+|r|+7}{4}} \exp(-w_*(\log \varepsilon)\, \varepsilon^{-1/4})$$

whenever $(q, \phi) \in [q_1, q_2] \times \mathbb{T}^2$, $\varepsilon \in]0, \varepsilon_0]$, $\mu \in [-\mu_0(\varepsilon), \mu_0(\varepsilon)]$, *where* $\mu_0(\varepsilon) = b^{-1} A(\varepsilon^{1/4}, \varepsilon^{1/4})^{-1} \varepsilon$.

Observe that the fixed width previously denoted by w_* is replaced by a new width $w_*(\log \varepsilon)$ which oscillates between two positive values, the new width being always larger than the constant value it previously assumed.

3.5.4. Moreover the upper bounds for the partial derivatives of the remainder $\Delta S - \mu \Delta S_1$ may now be compared to the lower bounds which are available for ΔS_1, at least when F satisfies some further assumptions (like in [DGJS]):

THEOREM. *Assume that*

$$F(q, \phi) = (\cos q - 1) f(\phi_1, \phi_2),$$

with an analytic function f *whose Fourier coefficients satisfy the following two conditions, for some constants* $K, a > 0$:
i) *For any* $k \in \mathbb{Z}^2$, $|f_k| \leq K\, e^{-h_0 |k|}$;
ii) *For any* $n \in \mathbb{Z}_+^*$, $|f_{\pm k^{(n)}}| \geq a\, e^{-h_0 |k^{(n)}|}$, *where we denote by* $k^{(n)} = (-\mathcal{F}_n, \mathcal{F}_{n-1})$ *the Fourier mode which corresponds to the Fibonacci sequence* $(\mathcal{F}_n)$.

Under these two assumptions, there exists a positive constant b *such that, for* $\varepsilon > 0$ *small enough and* $r_0 + |r| = 1$ *or* 2,

$$\max_{\phi \in \mathbb{T}^2} \left\{ |(\partial_q)^{r_0}(\partial_\phi)^r \Delta S_1(q, \phi; \varepsilon)| \right\} \geq b^{-1} \varepsilon^{-\frac{r_0+|r|+1}{4}} \exp(-w_*(\log \varepsilon)\, \varepsilon^{-1/4}).$$

Here Assumption i) simply says that f is analytic in a strip of width h_0 and bounded on its closure. As for Assumption ii) we first recall that the Fibonacci sequence is defined by: $\mathcal{F}_0 = 1$, $\mathcal{F}_1 = 1$ and $\mathcal{F}_n = \mathcal{F}_{n-1} + \mathcal{F}_{n-2}$ for $n \geq 2$. Assumption ii) is then a no gap condition: We preclude the possibility that the potentially dominant mode may vanish (or even be too small). This is in the style of Problem 3.5.1 except that here we can actually tell in advance *which* modes will dominate, because a) the dimension is 2 and b) we know *all* the convergents of χ.

Note that the left-hand side in the above inequality does not depend on the variable q because of the invariance of ΔS_1 under D_0. The first hypothesis on the Fourier coefficients of f could be replaced by the condition $A(\delta, \sigma) = \operatorname{cst} \delta^{-2} \sigma^{-2}$, which is slightly weaker.

Thus Theorem 3.5.3 indicates a bound which is always available for the size of the Poincaré-Melnikov approximation, and under the assumptions of Theorem 3.5.4

this approximation is not "abnormally small": Not only the exponent is not larger than $1/4$ but also the width is not larger than $w_*(\log \varepsilon)$). In that case one can then choose $A(\varepsilon^{1/4}, \varepsilon^{1/4}) = \text{cst } \varepsilon^{-1}$ and the above inequalities provide a range of values of μ for which $\mu \Delta S_1$ is indeed the dominant part of ΔS: Namely this happens at least on some interval $|\mu| \leq \text{cst } \varepsilon^{7/2}$. We finally note that Theorem 3.5.4 constitutes a direct generalization of one of the main results of [DGJS], inasmuch as that paper is confined to the isochronous case.

3.5.5. We close this paragraph with a result on the determinant of Hessian matrices, thus directly pertaining to the analysis of the homoclinic splitting matrix.

NOTATION. For any $\rho, h > 0$ we denote by $s_{\rho,h}$ the continuous $(4 \log \chi)$-periodic function of the real variable x defined *via* the formula:

$$s_{\rho,h}(x) = C_{\rho,h} \chi^{3/2} \cosh\left(\left(\frac{1}{2} - \left\{\frac{x - x_{\rho,h}}{4 \log \chi}\right\}\right) \log \chi\right),$$

with the same numbers $C_{\rho,h}$ and $x_{\rho,h}$ as above (see the notation in §3.5.2). We set $s_* = s_{\pi/2, h_0}$ and $x_* = x_{\pi/2, h_0}$. For $\kappa \in]0, 1/4[$, we define:

$$L_\kappa = \left\{ x \in \mathbb{R} \mid \kappa \leq \left\{\frac{x - x_*}{4 \log \chi}\right\} \leq \frac{1}{2} - \kappa \right\}.$$

Notice that $\mathbb{R} \setminus L_\kappa$ is the union of the intervals of length $4\kappa \log \chi$ which are centered on the solutions x of the equations $w_*(x) = C_*$ or $w_*(x) = C_* \cosh \frac{\log \chi}{2}$, *i.e.* the minima and the maxima of w_*. Notice also that:

$$s_{\rho,h}(x) = C_{\rho,h} \left[\cosh\left(\left\{\frac{x - x_{\rho,h}}{4 \log \chi}\right\} \log \chi\right) + \cosh\left(\left(1 - \left\{\frac{x - x_{\rho,h}}{4 \log \chi}\right\}\right) \log \chi\right)\right],$$

hence $2w_{\rho,h}(x) \leq s_{\rho,h}(x)$. According to Theorem 3.5.3 the Hessian matrix of the restriction of ΔS to any section $\{q = q_*\}$ is exponentially small with a width at least $2w_*(\log \varepsilon)$. In fact the width can be enlarged to $s_*(\log \varepsilon)$, and under the hypotheses of Theorem 3.5.4 it will actually reach this value, provided at least we exclude some intervals for the parameter ε by requiring that $\log \varepsilon \in L_\kappa$. This yields the following result, whose proof appears in [Sa2].

THEOREM. *Let $q_* \in]0, 2\pi[$ and $\kappa \in]0, 1/4[$. Under the assumptions of Theorem 3.5.4, for $\varepsilon > 0$ small enough such that $\log \varepsilon \in L_\kappa$, and for $\mu \in \mathbb{R}$ with $|\mu| \leq \text{cst } \varepsilon^{13/4}$, the function*

$$\phi \in \mathbb{T}^2 \mapsto \Delta S(q_*, \phi; \mu, \varepsilon)$$

is a Morse function with exactly four distinct critical points.

The absolute value of the determinant of the Hessian matrix at any of these point is bounded from above and from below by expressions of the form:

$$\text{cst } |\mu|^2 \varepsilon^{-1/2} \exp(-s_*(\log \varepsilon) \varepsilon^{-1/4}).$$

We thus get a bound *from below* on the size of the determinant of the splitting matrix and in fact also on each of its two eigenvalues.

3.6. Some variants and possible generalizations

In this paragraph we list some variants and generalizations which as a rule we have not worked out. This will be indicated and we may of course grossly underestimate the amount of work which is needed in order to effectively realize any of the suggestions below. Some of these remarks are also relevant in the more general framework of Chapter 2, and we are actually also trying to poke here in the direction of unifying the methods of Chapters 2 and 3, as well as possibly having periodic orbits and simultaneous approximation come more visibly into play.

The first obvious track is to question the necessity of Assumptions 3.1.2, playing around with the parameters entering Hamiltonian (∗).

3.6.1. Action-dependent perturbations.

The first generalization we contemplate consists in taking a perturbation F which depends on (q, ϕ, p, I) (possibly also on ε and μ, but this is inessential) and not only on (q, ϕ). That is, it does *not* satisfy the first part of Assumption 3.1.2 i); yet we still assume of course that it can be analytically extended to some complex strip as in the second part of that same assumption. We also assume here that the torus remains invariant after perturbation, that is we assume that the analog of Assumption 3.1.2 ii) holds true. The possibility of relaxing that assumption is briefly discussed in §3.6.2 below.

We first note that it is not difficult to extend the Hamilton-Jacobi algorithm at a formal level, as in Proposition 3.2.3. There appear more complicated terms stemming from the Taylor expansion

$$F(q, \phi, 2\sin(q/2) + \delta p, \delta I) = \sum_{(r_0, r) \in \mathbb{Z}_+ \times \mathbb{Z}_+^d} \frac{1}{r_0! r!} (\partial_p^{r_0} \partial_I^r F)(q, \phi, 2\sin(q/2), 0) \delta p^{r_0} \delta I^r,$$

with standard notation for the multi-integers ($r! = r_1! \ldots r_d!$, $\partial_I^r = \partial_{I_1}^{r_1} \ldots \partial_{I_d}^{r_d}$). The proof of the analyticity of the generating functions $S^\pm$ and the study of the domains of analyticity (Proposition 3.2.2 and especially Theorem 3.3.3) would then have to be rewritten, which may be quite cumbersome. At least the definition of the characteristic vector field given above in §3.4.3 is still valid in that case.

Let us indicate the formulas which generalize those of §3.2.3 in this situation. As in §3.2.3 we use the notation

$$S^\pm = S_0 + \mu T^\pm, \qquad T^\pm = \sum_{n \geq 0} \mu^n S_{n+1}^\pm(q, \phi; \varepsilon),$$

but the Hamilton-Jacobi equation now yields

$$D_0 T = -F(q, \phi, 2\sin(q/2) + \mu \partial_q T, \mu \partial_\phi T) - \frac{1}{2} \mu [\alpha (\partial_\phi T)^2 + (\partial_q T)^2].$$

We still denote by D_0 the unperturbed characteristic vector field whose inverses E^+ and E^- relative to $\mathcal{B}^+$ and $\mathcal{B}^-$ were described in §§3.2.3–3.2.4. We easily obtain the new algorithm

$$\forall n \geq 1, \quad S_n^\pm = E^\pm U_n^\pm,$$

$U_1^+(q, \phi) = U_1^-(q, \phi) = -F(q, \phi, 2\sin(q/2), 0)$ and

$$\forall n \geq 1, \quad U_{n+1}^\pm = -V_{n+1}^\pm - \frac{1}{2} \sum_{n_1 + n_2 = n-1} [\alpha \partial_\phi S_{n_1+1}^\pm \cdot \partial_\phi S_{n_2+1}^\pm + \partial_q S_{n_1+1}^\pm \cdot \partial_q S_{n_2+1}^\pm].$$

The term $V_{n+1}^\pm$ vanished in the case where F depended only on q and ϕ. In the general case it is determined by $S_1^\pm, \ldots, S_n^\pm$ as follows:

$$V_{n+1}^\pm(q,\phi;\varepsilon) = \sum_{\substack{(r_0,r)\in\mathbb{Z}_+\times\mathbb{Z}_+^d,\, r_0+|r|\geq 1 \\ m\in\mathbb{Z}_+^{r_0+|r|},\, |m|=n-r_0-|r|}} \frac{1}{r_0!\,r!}(\partial_p^{r_0}\partial_I^r F)(q,\phi,2\sin(q/2),0)\{dS^\pm\}_m^{r_0,r}(q,\phi;\varepsilon),$$

where, for $r_0 \in \mathbb{Z}_+$, $r \in \mathbb{Z}_+^d$ and $m \in \mathbb{Z}_+^{r_0+b_1+\cdots+b_d}$,

$$\{dS^\pm\}_m^{r_0,r} = (\partial_q S_{m_1^{r_0}+1}^\pm \cdots \partial_q S_{m_{r_0}^{r_0}+1}^\pm)(\partial_{\phi_1} S_{m_1^{r_1}+1}^\pm \cdots \partial_{\phi_1} S_{m_{r_1}^{r_1}+1}^\pm)$$
$$\cdots (\partial_{\phi_d} S_{m_1^{r_d}+1}^\pm \cdots \partial_{\phi_d} S_{m_{r_d}^{r_d}+1}^\pm),$$

with $m^{r_0} \in \mathbb{Z}_+^{r_0}$, $m^{r_i} \in \mathbb{Z}_+^{r_i}$ uniquely determined by the decomposition

$$m = m_0^r m^{r_1} \ldots m^{r_d}.$$

As in §3.2.5 we can spell out the first two orders of the algorithm. We will do that in the variable u, thus we introduce the functions

$$\tilde{F}(u,\phi,p,I) = F(q_0(u),\phi,p,I), \quad \beta(u) = (\tfrac{dq_0}{du})^{-1} = \tfrac{1}{2}\cosh u,$$
$$\tilde{U}_1(u,\phi) = -\tilde{F}(u,\phi,\tfrac{2}{\cosh u},0).$$

The first-order approximation of $\tilde{S}^\pm(u,\phi;\varepsilon,\mu) = S^\pm(q_0(u),\phi;\varepsilon,\mu)$ is now $\tilde{S}_0 + \mu\tilde{S}_1^\pm$ with

$$\tilde{S}_1^\pm(u,\phi;\varepsilon) = -\int_0^{\pm\infty} \tilde{U}_1(u+\zeta,\phi+\varepsilon^{-1/2}\zeta\omega)\,d\zeta.$$

Hence the Poincaré-Melnikov integral

$$\Delta\tilde{S}_1(u,\phi;\varepsilon) = \int_{-\infty}^{+\infty} \tilde{F}(u+\zeta,\phi+\varepsilon^{-1/2}\zeta\omega,\tfrac{2}{\cosh(u+\zeta)},0)\,d\zeta.$$

For $n=2$, a direct application of the algorithm yields

$$\tilde{S}_2^\pm = -\tilde{E}^\pm\left(\tilde{V}_2^\pm + \tfrac{1}{2}[\alpha(\partial_\phi \tilde{S}_1^\pm)^2 + \beta^2(\partial_u \tilde{S}_1^\pm)^2]\right)$$

with

$$\tilde{V}_2^\pm(u,\phi) = \gamma^{(p)}(u,\phi)\partial_u \tilde{S}_1^\pm(u,\phi;\varepsilon) + \gamma^{(I)}(u,\phi)\cdot\partial_\phi \tilde{S}_1^\pm(u,\phi;\varepsilon),$$
$$\gamma^{(p)}(u,\phi) = \beta(u)(\partial_p \tilde{F})(u,\phi,\tfrac{2}{\cosh u},0),$$
$$\gamma^{(I)}(u,\phi) = (\partial_I \tilde{F})(u,\phi,\tfrac{2}{\cosh u},0).$$

But some elementary manipulations of integrals provide more interesting formulas: one can indeed obtain a function $\hat{S}_2(u,\phi;\varepsilon,\zeta)$ which allows to write

$$\tilde{S}_2^\pm(u,\phi;\varepsilon) = \int_0^{\pm\infty} \hat{S}_2(u,\phi;\varepsilon,\zeta)\,d\zeta, \quad \Delta\tilde{S}_2(u,\phi;\varepsilon) = \int_{-\infty}^{+\infty} \hat{S}_2(u,\phi;\varepsilon,\zeta)\,d\zeta.$$

Such integral representations make it quite easy to see the exponential smallness of the oscillating part of $\Delta \tilde{S}_1$ or $\Delta \tilde{S}_2$. Here is the definition of $\hat{S}_2$ which generalizes the one given at the end of §3.2.5:

$$\hat{S}_2(u,\phi;\varepsilon,\zeta) = \hat{\chi}^{(u)} + \hat{\chi}^{(\phi)} - \Gamma^{(p)}\partial_u \tilde{U}_1(u+\zeta, \phi+\varepsilon^{-1/2}\zeta\omega)$$
$$- \Gamma^{(I)} \cdot \partial_\phi \tilde{U}_1(u+\zeta, \phi+\varepsilon^{-1/2}\zeta\omega),$$

where

$$\hat{\chi}^{(u)}(u,\phi;\varepsilon,\zeta) = \partial_u \tilde{U}_1(u+\zeta, \phi+\varepsilon^{-1/2}\zeta\omega) \cdot$$
$$\cdot \int_0^\zeta (B(u+\zeta') - B(u))\partial_u \tilde{U}_1(u+\zeta', \phi+\varepsilon^{-1/2}\zeta'\omega)\,d\zeta',$$

with $B(u) = \int_0^u \beta^2(u')du' = \frac{1}{8}(u + \frac{\sinh 2u}{2})$, and

$$\hat{\chi}^{(\phi)}(u,\phi;\varepsilon,\zeta) = \alpha\, \partial_\phi \tilde{U}_1(u+\zeta, \phi+\varepsilon^{-1/2}\zeta\omega) \cdot \int_0^\zeta \zeta'\, \partial_\phi \tilde{U}_1(u+\zeta', \phi+\varepsilon^{-1/2}\zeta'\omega)\,d\zeta',$$

$$\Gamma^{(p)}(u,\phi;\varepsilon,\zeta) = \int_0^\zeta \gamma^{(p)}(u+\zeta', \phi+\varepsilon^{-1/2}\zeta'\omega)\,d\zeta',$$

$$\Gamma^{(I)}(u,\phi;\varepsilon,\zeta) = \int_0^\zeta \gamma^{(I)}(u+\zeta', \phi+\varepsilon^{-1/2}\zeta'\omega)\,d\zeta'$$

(we had $\Gamma^{(p)}$ and $\Gamma^{(I)}$ vanishing in the case where F depended only on (q,ϕ)).

But in order to see the exponential smallness of the oscillating part of the whole difference $\Delta S = S^+ - S^-$, it is again preferable to use the method of the characteristic vector field.

3.6.2. Incorporating KAM. One can also consider the case where the invariant torus does depend on ε and μ, *i.e.* try to relax Assumption 3.1.2 ii). A KAM-type result is then needed to start with, in order to locate an invariant hyperbolic torus whose invariant manifolds can then be studied. Whatever the value of the twist vector α, the appendix contains the necessary KAM-type results in the case when the perturbation F depends on the angles only, that is if Assumption 3.2.2 i) remains in force. In fact, from the appendix, one immediately determines to what extent Assumption 3.1.2 i) can in turn be relaxed, that is what kind of dependence on the action variables is permissible for the perturbation, in order for KAM theory to apply. The answer is that one can in principle accomodate a dependence of the form $F = F(q, \phi, p, I_1)$, where I_1 corresponds to the *nonzero* components of α. This assumption is optimal; if the perturbation depends on some action variables corresponding to vanishing components of α, the invariant torus may very well simply cease to exist and in fact this probably generically happens, although we have not proved this assertion. One can also consider the variant where $\alpha = \alpha(\varepsilon)$ and some of the components vanish for ε tending to 0. The answer, still contained in the appendix, is essentially the same.

In fact not only does one have to locate the invariant torus, but one should also obtain a localized normal form around it with the largest possible domain of analyticity for the parametrization of the torus. This already says that one has to use Kolmogorov's original method, as is done in the appendix. It also says that one *has* to relax Assumption 3.1.2 i) simultaneously with relaxing Assumption 3.1.2 ii), because (generically) action-dependent terms will appear in the perturbation after the KAM process, *i.e.* in the system which will serve as a starting point in the analysis of the invariant manifolds. So the natural stable assumptions are exactly those which we have described above. Examining Kolmogorov's algorithm, one quickly discovers that such is the form of the perturbation which remains stable under the successive normalizing transformations, and this is how the statements in the appendix were discovered.

In order to go on with the method, one would have to put the torus at the origin by a symplectic change of coordinates. Being isotropic, the invariant torus must admit a parametrization of the form

$$\begin{cases} q = Q(\phi; \varepsilon, \mu) \\ p = P(\phi; \varepsilon, \mu) \\ I = \delta(\varepsilon, \mu) - P(\phi; \varepsilon, \mu)\partial_\phi Q(\phi; \varepsilon, \mu) + \partial_\phi \mathcal{A}(\phi; \varepsilon, \mu) \end{cases}$$

where δ is a constant vector and $Q, P, \mathcal{A}$ are functions on the torus. Thus one can choose new coordinates (q^*, ϕ^*, p^*, I^*) defined by

$$\begin{cases} q = Q(\phi) + q^* \\ \phi = \phi^* \\ p = P(\phi) + p^* \\ I = \delta - P(\phi)\partial_\phi Q(\phi) + \partial_\phi \mathcal{A}(\phi) + q^*\partial_\phi P(\phi) - p^*\partial_\phi Q(\phi) + I^*. \end{cases}$$

Of course the change of coordinates depends on ε and μ, and it is symplectic since one can write for it a generating function

$$W_{\varepsilon,\mu}(q, \phi; p^*, I^*) = qp^* + \phi \cdot I^* + qP(\phi) - p^*Q(\phi) - PQ(\phi) + \delta \cdot \phi + \mathcal{A}(\phi).$$

Instead of running the algorithm of §3.6.1 or the theory of the previous paragraphs in the variables (q^*, ϕ^*, p^*, I^*), one can also keep the isotropic invariant torus where it is and look for its invariant manifolds as *non*-exact Lagrangian graphs:

$$\mathcal{W}^\pm : \begin{cases} p = \partial_q S^\pm(q, \phi; \varepsilon, \mu) \\ I = \delta(\varepsilon, \mu) + \partial_\phi S^\pm(q, \phi; \varepsilon, \mu). \end{cases}$$

We only mention that this is in principle feasible, and the characteristic vector field is still defined because the defect of exactness δ is the *same* for both invariant manifolds (see §§1.7.1 and 1.10.1). In any event, it transpires that realizing the variants mentioned in this subsection does actually require some hard work! We direct the reader to [RW3] for an original use of KAM theory in this setting.

3.6.3. Simple versus multiple resonance. We now go on to follow a different and possibly important track, the point being that it may turn out to be quite general and geometrically significant, as the comments below suggest. Again we will be brief and sketchy.

In the framework of Hamiltonian (*) and in a down-to-earth version it amounts to studying frequency vectors of the form $\omega = \omega(\varepsilon) = (\omega_1, \sqrt{\varepsilon}\omega_2)$, where ω_1 is an n-vector and ω_2 is an m-vector, with $0 \le m < d$ and $m + n = d$; we thus include the case $m = 0$ which has been tacitly assumed hitherto. Here we refer to the rescaled version of (*), as it appears in §3.1.3. The situation corresponds to a resonance of multiplicity $m + 1$; beware again of the shift in the value of the parameter m compared to that in Chapter 2, which is due to the fact that we are staying here in the strip corresponding to the guiding resonance $p = 0$ (using Chirikov's terminology). Of course multiple resonances are dense on a simple resonance surface, and the small divisors $\omega \cdot k$ deserve their name, as they correspond to the vicinity of a (at least) double resonance. Now we assume that ω_1 is τ_1-Diophantine and we are back to a situation akin to the one studied in Chapter 2, basically with the same notation; in particular ω_1 is to satisfy precisely (5) in §2.1.4.

We will not repeat in details the remarks above about Assumptions 3.1.2, many of which could be adapted to the problem considered here, because we would like to sketch the appearance of more essential phenomena. However we stress the extremely important point that here we are still looking at invariant tori of dimension $d = \ell - 1$ (not n), whereas in Chapter 2 we were investigating the multiple resonance itself, with tori of dimension n. In particular, if one has to locate the invariant tori *via* some KAM technique, it is necessary here to impose a Diophantine condition on the complete d-vector ω. To wit, one should assume that it is τ-Diophantine for some $\tau \geq d$, *i.e.* it satisfies $|\omega \cdot k| \geq \gamma\sqrt{\varepsilon}|k|^{1-\tau}$ for all $k \in \mathbb{Z}^d \setminus \{0\}$ and some fixed $\gamma > 0$ where the factor $\sqrt{\varepsilon}$ is of course indispensible. Proposition A.1.4 in the appendix (replace ℓ with d there) discusses the amount of m-vectors ω_2 which are available, given a τ_1-Diophantine n-vector ω_1.

We come to the real point of this subsection, which is to try and make the connection with the results in Chapter 2, at least at an informal level; we will offer no proof of the facts listed below. Recall that the homoclinic splitting matrix always has size $d = \ell - 1$, after taking a section, whatever the multiplicity of the resonance and thus the dimension of the torus whose invariant manifolds intersect; in other words these are always Lagrangian, that is isotropic of maximal dimension ℓ. Now we simply contend that under the conditions above, the splitting matrix (again after taking a section) associated to the d-torus with frequency $\omega = \omega(\varepsilon)$ assumes exactly the *same* form (asymptotically as ε goes to 0) as the one displayed in Proposition 2.4.8 (with the only difference that m there is $m + 1$ here). In particular, one gets a block matrix with at least n exponentially small eigenvalues, with exponents at least $1/(2\tau_1)$. Moreover these n eigenvalues are supplemented with m eigenvalues which generically are *not* exponentially small. Although again we are looking at a *simple* resonance, we see that part of the splitting matrix approaches the transverse splitting matrix for the *multiple* resonance we are close to, which in our case is simply given by $p = I_2 = 0$. We insist that although this picture looks plausible and tantalizing we describe it informally without proof and have not tried to manufacture one.

We now note that the case $n = 1$ ($m = d - 1$), corresponding to a resonance of maximal multiplicity $\ell - 1$, or else to the case of unperturbed periodic orbits, is quite special. Indeed this is the *only* multidimensional case in which *no small divisors occur* (one may take $\tau_1 = 1$) and it can in fact be seen essentially as a parametrized version of the case with just *one* fast frequency, namely precisely the *scalar* ω_1. Perhaps one should rather say that it is actually one-dimensional as far as *singular* perturbation theory is concerned: There is just 1 $(= n)$ exponentially small eigenvalue, with exponent $1/2$ $(= 1/(2\tau_1))$. This all is very reminiscent of what happens with normal forms and which prompted the introduction of simultaneous approximation. And it would certainly be extremely interesting to study the *limiting process* which underlies the picture sketched above. In the case of normal forms, the limiting process goes along with the use of the Dirichlet approximation theorem and the reasoning is quite simple (see [L2] or [L3], p. 891) but also quite powerful, as for instance it also extends to infinite-dimensional problems (see §2.6.5 for references) where linear approximation simply breaks down. Here the limiting process seems to be more subtle but unraveling it would be extremely precious, making it in principle possible to simply bypass Problem 3.5.1, which looks like a serious stumbling block when $\ell > 3$.

3.6.4. In §3.4, the remainder $\Delta S - \mu \Delta S_1$ was particularly studied because of the hope that the Poincaré-Melnikov approximation would provide the leading asymptotic behaviour. This hope materialized in §3.5, albeit in very particular cases. However, were the first-order approximation abnormally small (*e.g.* because the right harmonics are lacking in the perturbation F), we could still try our luck with the second-order approximation, or more generally look for the first finite-order approximation of ΔS of the "right" exponentially small size. The Hamilton-Jacobi algorithm in principle enables us to compute the functions $S^{\pm}$ to any finite order with respect to μ and the methods of §§3.4–3.5 could in principle be adapted to bound the partial derivatives of the remainder

$$\Delta S - \mu \Delta S_1 - \cdots - \mu^n \Delta S_n$$

at any order n. Of course this strategy in no way bypasses Problem 3.5.1 and there would remain the problem of bounding from below the derivatives of $\Delta S_n = S_n^+ - S_n^-$.

Along this line one may also suspect that it would be better to perform a certain finite number of normalizing steps in order to "fish out" the important dominant Fourier modes of the perturbation (see [S] for similar ideas; see also [G1] in the case of just one fast frequency). But then why not normalize further? And by doing so one would return to ... the symplectic method implemented in Chapter 2. So this goes full circle between the symplectic and the analytic method and it is really not clear how these could be mixed in an efficient way.

At this point a side remark is in order, which connects to the last subsection. Assume that the perturbation is a trigonometric *polynomial* in the angle variables ϕ. Then the linear approximation says that the exponent of, say, the determinant of the splitting matrix is equal to $1/2$. Yet this is probably generically (with respect to the coefficients of the polynomial perturbation) simply wrong. The upshot is that *the prediction of the exponent via the Poincaré-Melnikov approximation is generically wrong for polynomial perturbations* except *in the absence of small divisors*. The last proviso concerns in particular the case $n = 1$ discussed in §3.6.3 above, for *any* value of d. Then of course the predicted exponent is just $1/2$ and this is the right answer, just as in the one-frequency case, *i.e.* the case $n = d = 1$.

3.6.5. We close this section with a short comment about the recent paper [GGM1], as it fits into the framework discussed above but with more concern about effectivity. In our notation, the paper deals with a subcase of Hamiltonian $(*)$ (still in the rescaled version of §3.1.3) corresponding to $d = 2$, $\omega = (\omega_1, \varepsilon^{1/2+a}\omega_2)$, $\alpha = (0, \varepsilon^{2a}\alpha_2)$, with $a \geq 0$. Here the quantities ω_1, ω_2 and α_2 are of course scalars. The perturbation $F = F(q, \phi)$ is supposed to be an even trigonometric *polynomial*. The paradigmatic cases are $a = 0$ and $a = 1/2$. The case $a = 0$ features in fact a *two*-timescale system, as in §3.6.3 above; the case $a = 1/2$ gives its title to the paper and was introduced, as the authors remark in their §2 (Remark (1)), because it naturally arises in a simplified model from celestial mechanics. In fact there is no serious technical difference between these two cases or with any value of $a \geq 0$. Picking $a > 0$ corresponds to moving closer to the double resonance $I_2 = p = 0$, but the critical value is actually $a = 0$. This is nothing but the well-known fact that resonances have natural width $\sqrt{\varepsilon}$, resulting in the factor $\varepsilon^{-1/2}$ appearing in front of Hamiltonian $(*)$ in §3.1.3.

The paper discusses the *determinant* of the 2×2 splitting matrix and especially the comparison of the linear and nonlinear values, *i.e.* the "justification of the Poincaré-Melnikov computation". The isochronous case $\alpha_2 = 0$ is studied in details, whereas the substantial modifications that are needed in the anisochronous (but not fully nonlinear) case $\alpha_2 \neq 0$ are sketched (see §7 and Appendix A.2). Clearly the crux of the matter is that we are actually in a case with just *one* fast frequency ($n = 1$), so that *no small divisors* can possibly show up. This explains why the linear computation is substantially correct for a polynomial perturbation; as for the exponent it is just equal to $1/2$, unless of course the Poincaré-Melnikov integral vanishes.

For applications of the "direct" method to two-timescale systems, still with a polynomial perturbation, we refer the interested reader to [Ga2], and to [GGM2] for the case where one lets the number of harmonics of the trigonometric polynomial tend to infinity as ε decreases.

3.7. A short historical tour and some concluding remarks

In this closing section we will again take the reader on a short historical tour, although with a purpose and flavour which markedly differ from those of the excursion at the end of Chapter 2. First the tour is shorter, more technical and the scene is much less wide. The "splitting problems" have grown into a rather technical and pointed subject. It can be considered either for its own sake or in connection with the general theory of near-integrable systems (not necessarily Hamiltonian), or as a branch of singular perturbation theory. We are in principle especially concerned with the second interpretation and have tried to present part of the context in §2.6, although much remains of course to be said. Here we will rather take the problem at its face value, without too much context, concentrating on the technical solutions which have emerged. (We leave aside the so-called direct methods, for which the reader may consult the articles by G.Gallavotti and coworkers.) This will hopefully help to clarify to some extent the relations between the various available methods and will naturally lead us to some concluding but also prospective remarks.

3.7.1. We proceed in chronological order, unavoidably starting with H.Poincaré. In fact many, if not most papers in the subject contain one or a few sentences quoting the Memoir on the three-body problem or the *Méthodes Nouvelles*, paying due but usually rather abstract respect to their author, who is essentially credited with having spotted the important problems. The reality is quite different and infinitely richer and the introductory remarks which follow are really just meant to tempt the reader. They are also justified because they connect quite directly with this chapter (see below). The splitting problem occurs essentially in two places in [P], at the end of the second and third volumes, both times as part of more general studies of homoclinic and heteroclinic solutions and with a (rather implicit) view toward the three-body problem. We will concentrate on the first passage (§§225 *sqq.*), which is of a more analytic nature. The second part (Chapter XXXIII) is more geometric and contains the celebrated sentence on the impossibility of drawing what is now called the homoclinic tangle (§397). The connection between "analysis" and "geometry" forms the subject of the short Paragraphs 401 and 402.

Right at the beginning of §225, Poincaré writes down a Hamiltonian which is the direct ancestor of (∗). Only problems with *one* fast frequency occur in [P] in this context although of course multifrequency problems are considered from other viewpoints. In any case the study of the problem of the "solutions doublement asymptotiques" is restricted to the planar case. Poincaré's model is thus essentially (∗) for $d = 1$. He does not assume that the unperturbed torus (*i.e.* periodic orbit) is invariant. To make clear the divergence of certain series which arise in this context, he introduces two parameters ε and μ. His model was generalized in [A1], also switching the roles of ε and μ, simply because in modern notation ε is the paradigmatic small parameter. Arnold's notation was subsequently adopted. No small divisors occur in [P] because it is a *one*-frequency problem. Arnold was concerned with geometry and does not really consider the analytic problem. He simply computes the linearized splitting, in which again no small divisors occur, this time because the perturbation is *polynomial*. Small divisors were added in [L1], leading to (∗) which is thus a very direct descendent of Poincaré's model.

The very next thing Poincaré does is to write down the Hamilton-Jacobi equation. This is completely natural to him and indeed it is one of the main tools in the *Méthodes Nouvelles*, including naturally for multidimensional problems. He then immediately expands the generating function in powers of ε (which is our μ) and writes down what we have called here the Hamilton-Jacobi algorithm. We have basically followed the same notation (see §225 Formula (7) and the discussion below it). Since the periodic orbit may be shifted he adds an ε-dependent constant. Poincaré is *a priori* interested in a "complete integral" of the Hamilton-Jacobi equation and indeed he does study the rotation and the libration, calling them the "ordinary cases".

Yet he says (opening sentence of §226) that: "Le cas limite où $h = 0$ présente plus d'intérêt". So he goes on to study this "limiting case" which includes the splitting problem. He first introduces the time along the unperturbed separatrix and calls it u; we follow his notation but he actually does it in a more general context, namely for any energy h, with the help of elliptic integrals. Then still in §226, he extends all the variables to complex values, somewhat implicitly though, because it is the obvious thing to do and need not really to be mentioned. He then performs computations with Cauchy integrals which make the splitting appear as a residue. At the beginning of §227 he adds that this could (and will) be done in a simpler way but that these residue computations actually show how to understand the transition from the "ordinary case" ($h \neq 0$) of libration or rotation to the "limiting case" ($h = 0$), which he proceeds to briefly explain using the degeneracy of elliptic functions from doubly to simply periodic functions.

In §228 he develops the "simpler" way alluded to in §227, shows that the solutions of the Hamilton-Jacobi equation can be formally expanded in powers of $\sqrt{\mu}$ (our $\sqrt{\varepsilon}$) but that the expansion is divergent, because of the existence of a nonzero splitting: The two solutions of the Hamilton-Jacobi do not coincide, yet their formal expansions *do* coincide and thus cannot be convergent. He then discusses the double expansions in ε and $\sqrt{\mu}$ for any energy h and displays the causes of divergence, *i.e.* what he calls the "large multipliers". He actually in essence proves that one is dealing with what we would call Gevrey-1 expansions, the size of the general coefficient of the typical power series going roughly like $n!$.

This completes our terse summary of the analytic part of Poincaré's study. It goes without saying that the "Poincaré-Melnikov" integral is not mentioned as such but occurs as a natural, if not obvious first-order computation. Of course it was even more natural to Melnikov over sixty years later (1963) and the latter author was actually investigating different and more subtle questions.

Around 1990 one of us (P.L.) noted the wealth of actual and potential results contained in this half-forgotten study by Poincaré, as well as the fact that it seemed to contain the germ of a resurgent (in the sense of J.Écalle) treatment of the problem. This was subsequently made quite tangible by another of the authors (D.S.) in [Sa1], to which we refer for further information. We simply mention that resurgence theory is concerned with the *global* analytic structure of the problem and that the one-frequency case seems to give enough food for thought at the moment (see however [Sa2], Remark 3.1). At any rate that experience led us to use the Hamilton-Jacobi formalism in the multifrequency case as well. Another ingredient of the analytic method used in this chapter came from the next story.

3.7.2. What we called the *analytic method* was invented and developed by V.F.Lazutkin, starting with [La1]. For the sake of (somewhat more than) the anecdote one may point out that the conditions of (non) publication of that paper seemed to be designed to ensure an absolute minimal readership. Written during the Brezhnev era, at a time where samizdat (literally "self-publication") was common, one may recall that in the first place VINITI is not by any means a journal or a publishing house but an office with a view to ensure a minimal copyright (intellectual, not commercial it goes without saying) to the authors of the manuscripts deposited there. Next [La1] is naturally written in Russian and has never been translated. Last but not least the paper attacked in fact *several* problems at one go. The point is that the splitting problem for the standard map is substantially more difficult than the analogous problem for, say, the rapidly forced pendulum; and for several reasons, which are well-known. For instance it is a discrete problem and it is nontrivial to analyse even the simplest linear *difference* operator (see [La1], §6). In any case the paper was evidently confined at first to a small circle around the author. Yet it quickly benefited from the coming historical changes and it took only a few years for at least part of its content to filter out and around. The delay was indeed much shorter than for *e.g.* N.N.Nekhoroshev's work dating from 1970, a phenomenon which again can at least partly be ascribed to historical circumstances. Like many seminal papers [La1] contains some moot points, loose ends and even some incorrect statements. It took fifteen years and quite a few papers to clarify, tie up and correct these. We mention [LST], [La2] and [GLS] but this represents by no means a complete list. In fact [G2] may feature the last episode in that story, and we refer the reader to that text for more information and a bibliography.

So V.F.Lazutkin introduced the analytic method in [La1], in the one-frequency case but in a context which is more involved than that of the continuous models which were subsequently studied. He emphasized the simple and all important idea (which apparently does *not* appear in [P] ...) of getting domains of analyticity which are *as large as possible* in order to get a good control of the functions on the real domain, *via* a fairly simple lemma, of which Lemma 3.4.2 is the direct descendent. This has been used in many papers since and is the second main ingredient in this chapter. One of the important messages of resurgence theory is that it is worth studying the analytic structure of the problems from an even more global viewpoint, but again even the one-frequency case is far from completely understood at present.

3.7.3. The next and last paper we will briefly discuss is [HMS], which was published a few years after [La1] but essentially independently, despite a passing reference to [LST] as a preprint. The simplest model problem which is studied there is the periodically forced pendulum with fast forcing frequency, which is a symplectic but not exact system. The authors develop a variant of the analytic method, although without the help of Lazutkin's lemma on Fourier coefficients. They use instead a classical iteration method in suitable Banach spaces of analytic functions and do stress the importance of getting wide analyticity strips; the proofs are only sketched and were apparently never fully written out by the authors but the sketch they provide is both sufficiently conceptual and detailed to be convincing.

The point we want to make is quite simple indeed: the paper ([HMS]) is *not* written from a Hamiltonian viewpoint. And it seems this had an almost fortuitous but quite serious influence on the recent history of the subject. That the paper is not written in a Hamiltonian context has a rather obvious cause: the authors have certain non-Hamiltonian applications in mind. They start (see their §3) from a general planar system of differential equations, assuming only a reversibility property for convenience. They view the splitting problem as part of singular perturbation theory, having to do with what is sometimes termed "asymptotics beyond all orders", and the reader will find in their §6 an interesting list of motivating problems with references, some of which are Hamiltonian and others not; of course these are all problems with only one fast frequency (or the like). Since the setup is not *a priori* Hamiltonian, the iteration method (§§3–4) naturally is not either. In fact even in the framework of *regular* perturbation theory it turned out to be a nontrivial and indeed rather cumbersome task to develop general (non-Hamiltonian) schemes which compute "splitting-like" quantities order by order in any dimension. This is done in [HMS] in the planar case, although the emphasis there is naturally on *singular* perturbation theory, *via* complexification of the variables and a variant of the analytic method.

There soon followed a flowering of papers, some of which are based, more or less loosely, on the technique inaugurated in [La1], others on [HMS]. The point was and still is to explore all the ramifications of the one-frequency case, which is far from completely understood from a global viewpoint. One can again compare with the *linear* one-dimensional WKB method, whose global version (pioneered in particular by A.Voros) gave rise to what is called "quantum resurgence". We refer to [DS] for a survey of the story, as well as to [Sa1]; both papers contain the necessary references although naturally their lists cannot possibly be exhaustive. In fact we would like to add [EKS], which was published after [DS] and appears in [Sa1] only as a preprint.

All the papers we have mentioned or hinted at until now deal with systems with only one fast frequency and are in fact almost always planar Hamiltonian systems. Under such circumstances it does not make very much difference whether one uses the method of [HMS] (or a variant thereof) which was designed to accomodate non-Hamiltonian systems, or a "truly" Hamiltonian method. In other words one can iterate the variational equation without making use of the Hamilton-Jacobi equation. If the system is in fact Hamiltonian, one simply has to manipulate *two* scalar quantities instead of one, but even that disappears if one for instance assumes that the time-dependent Hamiltonian is even with respect to the time variable. However as soon as one goes to multidimensional dimensional Hamiltonian systems, the difference becomes enormous. Symplecticity is not just volume preservation any more; it involves a lot of symmetry! And any insight into this symmetry is simply lost by writing out the variational equation, even for a system as simple-looking as the one associated with Hamiltonian ($*$). Instead of manipulating the scalar generating functions $S^\pm$, one has to deal with vectors, which are not easily recognized as the gradients they actually are. In fact fiery discussions flared about the possible symmetry of the splitting matrix which is now recognized as the Hessian matrix of ΔS. To summarize our point again, it seems that the scheme developed in [HMS] led many of its direct or indirect readers into the treatment of Hamiltonian problems in a non-Hamiltonian way. This was not very consequential in the one-frequency case but turned out to be quite awkward as soon as one starts studying the multifrequency (or multidimensional) case. In the present paper on the contrary, we have tried to stay as much as possible within the Hamiltonian framework, making use of all its specificities. An obvious negative corollary is that *the present paper says nothing in the non-Hamiltonian case*. We also mention as a last fact that we do not know to-date of a single study of a true singular multifrequency non-Hamiltonian system. Although this may sound like a long list of qualifying adjectives, it plainly comprises in fact an enormous number of "natural" systems (see [LM] for the counterpart in terms of normal forms and averaging).

3.7.4. In this final subsection we quickly review the content of the paper, with an eye on problems and possible developments. Chapter 1 above is concerned with fairly general geometric problems, which means that a large part of the description is not confined to the perturbative setting. Yet it is essentially symplectic. We feel it useful to develop the picture beyond our immediate needs for at least two reasons. First it gives a hopefully sound and firm geometric basis to the (Hamiltonian part of the) subject. Second it is liable to be useful for attacking other problems. For instance, staying in the perturbative framework, the splitting problems form only a small part of the general study of the global instability in near-integrable analytic Hamiltonian systems. This may not even be an unavoidable ingredient, if one is able to push beyond the original Arnold mechanism and variants thereof. Chapter 1 hopefully provides some fundations and intuitions, including in the direction of comparing and merging geometric and variational methods.

We have found no predecessor (at least in print) for the results detailed in Chapter 2, although with hindsight the method seems to us to be a natural application of resonant normal forms to an exponentially high order, combined with some stability statements from the perturbation theory of self-adjoint operators. The geometric picture which emerges, as embodied especially in Theorem 2.4.9, essentially fits with the corresponding picture coming from long time stability theory, as explained in §2.6.6. It is also quite general, dealing essentially with any analytic near-integrable Hamiltonian system and with resonances of any multiplicity. One gets a bound from above for the splitting matrix in a very detailed form, *i.e.* for the individual eigenvalues. This bound is probably optimal as far as the exponents are concerned. But ... there seems to be little hope to be able to actually prove this optimality by simply refining the symplectic method of Chapter 2; we briefly return to this point below.

Chapter 3 digs deeper into the analytic structure of the problem, using three main ingredients. First the Hamilton-Jacobi equation which as we have seen looks like the natural tool to use, given the circumstances; at least so it appeared to Poincaré a century ago. Second we use the analytic method as inaugurated in [La1] (and [HMS]). It is also applied in [DGJS] and [RW1] which to our knowledge are the only two papers (prior to the present one of course) dealing with the splitting problem for multifrequency systems and confronting of course the specific difficulty of the emerging small divisors. Yet we have seen that [DGJS] studies a very specific *isochronous* problem, whereas [RW1] is not errorfree, although it may be amended for the most part (see also [RW3]).

The setting in Chapter 3 is clearly narrower than that of Chapter 2, but apart from getting a better insight into the analytic structure of the problem the main gain consists in comparing the full splitting with the linearized splitting, as in particular in Theorem 3.4.6. This required adding a third and apparently new ingredient, namely the specific use of the characteristic vector fields. As we already pointed out, this may depart slightly from the fully symplectic philosophy, or, say, lie more on the Lagrangian rather than the Hamiltonian side.

Now in order to exploit the estimate of the difference between the full and the linearized splittings, there remains to evaluate the latter. And this is where the prospects look rather grim. Another way to put it is that Problem 3.5.1 looks extremely hard if $\ell > 3$. Of course such assessments do not mean much but at any rate solving this or similar problems seems like an unavoidable prerequisite if one wants to continue along this path. The exponents found in Chapter 3 coincide with particular cases of those in Chapter 2 and, to assert that these values are indeed the right ones, one simply needs a precise estimate of the linearized splitting, which comes down to Problem 3.5.1, or in fact somewhat more. In the particular case of systems with three degrees of freedom ($\ell = 3$) one reduces the arithmetic to scalars. The problem can then be solved fairly easily for the golden mean, and probably also for any quadratic irrational. We have borrowed the solution for the golden mean from [DGJS] and used it in §3.5, leading to Theorems 3.5.4 and 3.5.5, whose statements and proofs are already quite involved. In principle §6.3 of [RW1] and [RW2] should provide the possibility of a full treatment of the case $\ell = 3$, producing the analogs of Theorems 3.5.4 and 3.5.5, at least for numbers of constant type. However the higher-dimensional cases $\ell > 3$ again seem to be wide open and not easily accessible.

Faced with this situation, we mention in closing some as yet very uncertain tracks. One could try and unify the approaches in Chapters 2 and 3, *i.e.* the symplectic and analytic methods. This was briefly discussed in §3.6.4 above and may be recast in the following way: in the symplectic method, the splitting shows through in the divergence of the sequence of normalizing transformations; in the analytic method it manifests itself through the analytic structure, in particular that of the unperturbed homoclinic trajectory and of the perturbative term in the Hamiltonian function. Can one understand the connection better? This certainly sounds like an interesting program *per se* but it is not clear whether one will not in the end stumble again over Problem 3.5.1, as far as "evaluating the splitting" is concerned.

Another possibility (or impossibility) which was already hinted at, in particular in §3.6.3, consists in trying to just go around the difficulty, introducing simultaneous approximation, that is considering multifrequency problems as limiting cases of one-frequency problems. Here it is rather the problem of going to the limit which seems quite subtle, but this strategy, which was quite successful in the study of long time stability, at least makes it in principle possible to go *around* Problem 3.5.1 and it is also connected with more geometric considerations. Time will tell.

APPENDIX

Invariant Tori With Vanishing or Zero Torsion

In this article we investigate not only the fully nonlinear case, but also, to some extent, mixed cases where torsion (also called twist) is degenerate, or vanishing, *i.e.* goes to zero together with the appropriate parameter. In the model Hamiltonian (∗) of the introduction, this means that some components of α may be identically zero or that α depends on a small parameter *e.g.* $\alpha = \alpha(\varepsilon)$ and some components vanish as $\varepsilon \to 0$. It turns out that under certain conditions the limit of vanishing torsion is *regular*, regarding in particular the existence of invariant tori. This provides in particular a partial bridge between the isochronous and anisochronous cases.

In this appendix we have gathered some KAM-type results covering these cases, in a way which seems to be close to optimal (in a loose sense of the word). Results of this type were first derived by G.Gallavotti and coworkers in a more restricted setting (*cf.* [Ga1,2], [Ga-Ge]) using the so-called "direct methods"; here we follow [L6]. The proofs are essentially complete in the elliptic case (§§A.1.1–A.1.4) but we confine ourselves to short indications in §A.1.5 concerning the routine extension to the 1-hyperbolic case. In fact, once it is understood that these results lie fully within the range of traditional methods and indeed require only a minor modification of the original scheme proposed by Kolmogorov, the whole toolbox of classical KAM theory becomes available, so that dozens of variants, improvements etc. are available for free.

To get a feeling of what is happening, let us start with a trivial situation. Consider the Hamiltonian: $H(I,\phi) = \omega \cdot I + \varepsilon f(I,\phi)$, with $(I,\phi) \in \mathbb{R}^\ell \times \mathbb{T}^\ell$ ($\ell > 1$). In general H generates a very nontrivial flow (and nontrivial problems!). But assume that in addition $f = f(\phi)$ is in fact independent of I. Then writing out the equations of the motion:

$$\frac{d\phi}{dt} = \omega, \quad \frac{dI}{dt} = -\varepsilon \frac{\partial f}{\partial \phi},$$

one sees that the system is trivially integrable (using one "quadrature" *i.e.* integration) and that all solutions have (quasi)frequency ω. Here this is not even a perturbative statement (set $\varepsilon = 1$) but it leads one to suspect that there are cases when quasi-periodic solutions persist independently of the nonlinearity, and that this is connected with the absence of action variables in the perturbation (in essence the above remark is due to G.Gallavotti). We now move to precise statements.

A.1.1. We start with the elliptic case *i.e.* the case of tori of maximal dimension. The results below follow from a minor variation of the scheme originally proposed by A.N.Kolmogorov in [K]. We basically need only to refer to [K] and its nice pedagogical elucidation in [BGGS]; we use Lie transforms as in [BGGS], as they appear to be slightly more convenient than the canonical transform formalism of [K]. Although we will be sketchy in the exposition, the proofs in the elliptic case below are complete, granted only [K] with the estimates spelled out as in [BGGS].

For the convenience of the reader, we have tried to make this appendix essentially self-contained, so that we recall first a few pieces of notation, which depart only slightly from those used in the rest of the text. So let $(I,\phi) \in \mathbb{R}^\ell \times \mathbb{T}^\ell$ be the phase variables, and split ℓ as $\ell = n + m$. Write (I_k, ϕ_k) ($k = 1, 2$), $I = (I_1, I_2) \in \mathbb{R}^n \times \mathbb{R}^m$, $\phi = (\phi_1, \phi_2) \in \mathbb{T}^n \times \mathbb{T}^m$, and similarly for the other functions on phase space (I, ϕ). We use a rather concise notation for these tensorial (scalars and matrices in fact) quantities but it should cause no ambiguities. For example, if ω is a frequency and I an action, we write ωI or $\omega \cdot I$ for the scalar product of ω and I; in the same vein, CI^2 should be read as $CI \cdot I$, with C a square matrix. The first result reads as follows:

THEOREM. *Consider the nearly integrable Hamiltonian:*

$$H(I,\phi) = \omega I + \frac{1}{2}C_1 I_1^2 + \frac{1}{2}\tau C_2 I_2^2 + \varepsilon f(I_1, \phi) + \varepsilon\tau g(I, \phi),$$

where ε and τ are real parameters. Assume that the symmetric matrices C_1 and C_2 are invertible, that the functions f and g are defined and analytic near $I = 0$, and that the vector ω is Diophantine.

Then for $|\tau| \leq 1$ and $|\varepsilon| \leq \varepsilon_0$ with $\varepsilon_0 > 0$ independent of τ, there exists an invariant torus for H which is ε-close to $I = 0$ (uniformly in τ) and on which the flow is conjugate to the rotation with frequency ω.

There are several variants of the statement which could be proved in the same way. Here f might depend on ε and g on ε and τ, in which case one requires analyticity in ε (but not necessarily in τ). We do not assume that C_1 and C_2 are diagonal, simply because it would not lead to any simplification. We could in fact consider more generally a Hamiltonian $H(I, \phi, \varepsilon, \tau)$ analytic in (I, ϕ, ε) and, say, C^2 in τ, and give the appropriate conditions on its partial linearization at $\tau = 0$ which would lead to a seemingly more general statement than the one above and would recover the original statement of Kolmogorov in [K] for fixed nonzero τ. In any case, for fixed $\tau \neq 0$, this theorem is just a usual KAM-type statement. Now for $\tau = 0$, one recovers the torsionfree case, and since f does not depend on I_2, one is dealing with the ω_2-quasiperiodic perturbation of a fully nonlinear integrable Hamiltonian with n_1 degrees of freedom. The statement says that the limit $\tau \to 0$ is actually *regular*, although the partial torsion or twist matrix τC_2 vanishes.

Another remark is that we could consider τ as an m-vector, *i.e.* deal with "multiple actionscale" Hamiltonians, rather than the two actionscale case of the theorem. This amplification would come essentially from a proper reading of the formulas: τ would be an m-vector, τI_2 would mean componentwise multiplication (*not* scalar product), $\tau C_2 I_2^2$ would mean $(\tau I_2 \cdot C_2 I_2)$ etc. Then we could simply identify (some of) the eigenvalues of the torsion matrix (the vector α in our basic model Hamiltonian (*)) with the torsion *vector* τ. For simplicity, we will restrict to a scalar parameter τ (which of course has nothing to do with the Diophantine exponent; we apologize for this notational clash).

As mentioned above this kind of situations was first considered by G.Gallavotti and coworkers under the heading "twistless tori"; we prefer to call them tori with *vanishing* torsion or twist since the torsion is not necessarily zero but may vanish in some limit.

A.1.2. In order to prove Theorem A.1.1, one first recasts (just as in [K]) the Hamiltonian into "Kolmogorov's normal form". Indeed Theorem A.1.1 is an immediate consequence of the following

PROPOSITION. *Consider the Hamiltonian:*

$$H(I,\phi) = \omega I + \frac{1}{2}C_1(\phi)I_1^2 + \frac{1}{2}\tau C_2(\phi)I_2^2 + \varepsilon A(\phi) + \varepsilon B_1(\phi)I_1 \\ + \tau\varepsilon B_2(\phi)I_2 + Q(I_1,\phi) + \tau R(I,\phi),$$

where ω is Diophantine, the averages $\overline{C}_1$ and $\overline{C}_2$ of the symmetric matrices $C_1(\phi)$ and $C_2(\phi)$ over $\phi \in \mathbb{T}^\ell$ are invertible, $A(\phi)$ has zero average and Q (resp. R) is of order I_1^3 (resp. I^3). Here all the intervening functions are supposed to be defined and analytic in their respective arguments near $I = 0$. The conclusion is the same as in Theorem A.1.1.

REMARK. In both the above theorem and proposition, we chose (in contrast with [K] and [BGGS]) to make the dependence on the paramaters ε and τ explicit, simply because there are two of them, and we hope this is typographically helpful. Yet we have dropped this dependence from the functions and we hope that the reader will mentally restore it. In fact, since the proof is iterative, one can and needs to assume that the functions may depend on the parameters.

PROOF. As noted above, the proof of Proposition A.1.2 is almost identical to its analog in [K] (and [BGGS]). So we will be sketchy and mention only the few points of difference. First write $H = H^0 + \varepsilon H^1$ where $H^1 = A + B_1 I_1 + \tau B_2 I_2$ is the perturbation which we want to eliminate recursively. Let χ be the auxiliary Hamiltonian (the analog of the generating function in terms of Lie series) by means of which we will perform one step of the perturbation scheme. Here we choose (again essentially as in [K]):

$$\chi(I,\phi) = \xi \cdot \phi + X(\phi) + Y_1(\phi)I_1 + \tau Y_2(\phi)I_2,$$

where $\xi \in \mathbb{R}^\ell$, the scalar function X and the vector functions Y_1 and Y_2 (of sizes n and m respectively) have to be determined. Note that for $\tau \neq 0$ the I_2-component ξ_2 of the mean translation of the torus is *a priori* of order 1 (not τ).

We perform one step of the perturbation scheme; the convergence will follow exactly as in the usual case (*cf.* [BGGS] §5). Denote by L_a the Liouville operator associated to a function a; that is $L_a(b) = \{a,b\}$ is the Poisson bracket of a with a function b. We modify H by taking the time ε of the flow of χ; so we get the new Hamiltonian $H' = \exp(\varepsilon L_\chi)H$ and as usual we have to pick χ so as to make the perturbative term smaller. We keep the same names for the variables when referring to H' (this is one of the convenient features of Lie series; see *e.g.* [BGGS] §3.1) but the new functions will get "primed" names (A becomes A' etc.). Again here we closely follow §4 of [BGGS] which completely develops a few lines from [K].

The function χ is determined by solving the "homological equation", namely here by requiring that:

$$H^1 + \{\chi, H^0\} = cst + O(I_1^2) + \tau O(I^2).$$

So we compute the left-hand-side:

$$H^1 + \{\chi, H^0\} = -\xi \cdot \omega - \omega \frac{\partial X}{\partial \phi} + A(\phi)$$
$$+ [B_1(\phi) - C_1(\phi)(\xi_1 + \frac{\partial X}{\partial \phi_1}) - \omega \frac{\partial Y_1}{\partial \phi}]I_1$$
$$+ \tau[B_2(\phi) - C_2(\phi)(\xi_2 + \frac{\partial X}{\partial \phi_2}) - \omega \frac{\partial Y_2}{\partial \phi}]I_2 + O(I_1^2) + \tau O(I^2).$$

Here one should be a little careful when unravelling the notation. the keypoint is that in the last two lines, which essentially duplicate the usual result, one gets the *partial* ϕ-gradients of the scalar function $X(\phi)$ but the *total* gradients of the vector functions $Y_1(\phi)$ and $Y_2(\phi)$.

The system to be solved reads:

$$\begin{cases} \omega \frac{\partial X}{\partial \phi} = A(\phi) \\ \omega \frac{\partial Y_1}{\partial \phi} = B_1(\phi) - C_1(\phi)(\xi_1 + \frac{\partial X}{\partial \phi_1}) \\ \omega \frac{\partial Y_2}{\partial \phi} = B_2(\phi) - C_2(\phi)(\xi_2 + \frac{\partial X}{\partial \phi_2}). \end{cases}$$

The reader may want to contemplate this system for a minute, as its solvability is the keypoint of the whole thing. The rest goes through "as usual" and again scores of variants and improvements are then available on the market.

The first equation is the same as usual and can be solved because $A(\phi)$ has zero average and ω is Diophantine. The last two equations are identical and indeed exactly duplicate the usual one (see [K] or eq. (4.14) in [BGGS]). So they can be solved for the appropriate (and unique) choice of $\xi = (\xi_1, \xi_2)$, because the averages of $C_1(\phi)$ and $C_2(\phi)$ are invertible, and using again the fact that ω is Diophantine. Moreover, the estimates on ξ, X, Y_1 and Y_2 are clearly the same as usual, and for good reasons.

This essentially finishes the proof. We mention for completeness the last insignificant departure from the usual scheme that needs to be brought in the evaluation of the remainder $H' - H - \{\chi, H^0\}$. To this end, one has simply to estimate $\{\chi, H^1\}$ and $\{\chi, \{\chi, H\}\}$ (second order Taylor expansion; see [BGGS] §3.1). The minor point we want to make is that the τ-independent and the τ-dependent terms should not be lumped together. More precisely, one starts with estimates for A, B_1 and B_2 and wishes to derive similar estimates for A', B_1' and B_2'. As mentioned above, one first gets estimates for ξ, X, Y_1 and Y_2, the exact same as usual. Now decompose H^1 and χ as $H^1 = H_0^1 + \tau H_1^1$, $\chi = \chi_0 + \tau \chi_1$, with $H_1^1 = B_2 I_2$ and $\chi_1 = Y_2 I_2$; each of H_0^1, H_1^1, χ_0 and χ_1 has thus already been estimated. It is now immediate to separately estimate the order 0 (with respect ot τ) and order 1 parts of $\{\chi, H^1\}$ and $\{\chi, \{\chi, H\}\}$. Here H can be left as is and one does not even have to expand; just count the number of terms. This is a minor point, and the rest literally follows the standard case, finishing the proof of Proposition A.1.2 and thus of Theorem A.1.1. □

We add some brief remarks on well-known issues, namely nondegeneracy conditions, arithmetical conditions and smoothness conditions. These would also essentially apply to the hyperbolic case discussed below although most of them have not been implemented in literature, outside the case of tori of maximal dimension. Concerning the first of these topics, we required here that C_1 and C_2 be invertible, which is a condition on the second order jet of the unperturbed Hamiltonian. This can be considerably weakened, as in particular the work of H.Rüssmann demonstrates (see [R1,3]). In order to apply the results in the present setting, consider again a Hamiltonian $H(I, \phi, \varepsilon, \tau)$ which is nearly integrable (and integrated), *i.e.* $H(I, \phi, 0, \tau)$ is actually ϕ-independent. The nondegeneracy conditions will then be imposed on the ϕ-independent functions $H(I, \phi, 0, 0)$ and $\partial H/\partial \tau(I, \phi, 0, 0)$. Above we simply considered the Hessian matrices of these two functions. Coming to the question of the arithmetical conditions, one could probably prove the above under Bruno's arithmetic condition with no less and no more efforts than are required in the usual case (see [R2,3]). As for smoothness conditions, the usual remarks are in order and Theorem A.1.1 and Proposition A.1.2 (probably) hold for data of class C^r with r large enough.

A.1.3. It should be noted that Theorem A.1.1 has been stated in a somewhat particular setting, as mentioned after the statement, simply for the sake of simplicity. But Proposition A.1.2 is stated in a general way, which means that it exactly parallels the corresponding statement in [K] (*cf.* §2 in [BGGS]). As a consequence it immediately proves KAM-type results in a general setting. Let us quote one important example with *identically* zero torsion, namely that of the quasiperiodic perturbation of a fully nonlinear Hamiltonian. We have the following statement:

THEOREM. *Consider the non-autonomous Hamiltonian:*

$$H(I, \phi, t) = h(I) + \varepsilon f(I, \phi, \omega t),$$

where:
i) $(I, \phi) \in \mathbb{R}^n \times \mathbb{T}^n$ $(n \geq 1)$, $\omega \in \mathbb{R}^m$ *is Diophantine;*
ii) *the perturbation f is a real function on $\mathbb{R}^n \times \mathbb{T}^\ell$, $\ell = m + n$, and the whole Hamiltonian H is analytic over $D \times \mathbb{T}^\ell$ with D some domain in I-space;*
iii) *the twist matrix $\partial^2 h/\partial I^2$ is nondegenerate over D.*

Then, as ε tends to 0, there is a set of invariant tori of asymptotically full measure in $D \times \mathbb{T}^n$.

The conclusion of this theorem is stated somewhat informally and incompletely but it is actually the same as in the usual KAM-type statements. The invariant tori are close to the unperturbed ones, the flow on them is conjugate to a rotation etc. This result is an immediate consequence of Proposition A.1.2 with $\tau = 0$ and the reduction to that statement is as in [K]. Indeed one of the main technical inputs in Kolmorogov's paper consists in this normal form localized near a torus, now known as "Kolmogorov's normal form".

A.1.4. Perhaps the reader said: but ... because there remains in fact one moot point, namely the existence of many Diophantine vectors of the sort we need, so that in particular we do get a set of tori of asymptotically full measure, as asserted in the theorem. This is a point which is specific to the case with identically zero torsion and which does require a little arithmetic study, as we proceed to explain.

What we need to prove goes as follows. We are *given* a fixed m-dimensional Diophantine vector ω and we wish to estimate the measure in (the unit cube of) $\mathbb{R}^n$ of the vectors ω' such that the concatenation (ω, ω') is a Diophantine ℓ-vector. The answer is given in the following proposition, which although comparatively elementary, does not seem to appear in the literature on Diophantine approximation. It is a pleasure to thank V.Bernik for help with the proof of this proposition which gives quantitative information on the relative measure of the invariant tori in Theorem A.1.3 above and other similar statements.

We start afresh with a slightly different notation for convenience. So let $\alpha \in \mathbb{R}^m$ be a fixed τ_1-Diophantine m-vector, more precisely: $|\alpha \cdot a| \geq \rho |a|^{1-\tau_1}$ for $a \in \mathbb{Z}^m \setminus \{0\}$ and some $\rho > 0$, $\tau_1 \geq \ell$. Here and below, we use the supnorm, as is usual in approximation theory. Namely we put $|a| = \sup_j |a_j|$, where the a_j are the components of a. Let $n \geq 1$ be a positive integer, and write $\ell = m + n$. We now have the following

PROPOSITION. *Let $\alpha \in \mathbb{R}^m$ be τ_1-Diophantine in the sense recalled above, and let τ be real with $\tau > \tau_0 = \sup(\ell, \tau_1)$. Let $\beta \in \mathbb{R}^n$ and write $\gamma = (\alpha, \beta)$. Then for Lebesgue almost all β, the inequality $|\gamma \cdot c| < |c|^{1-\tau}$ has only a finite number of solutions in integer vectors $c \in \mathbb{Z}^n$.*

Thus the set of points β such that $\gamma = (\alpha, \beta)$ is a τ-Diophantine ℓ-vector has full measure in $\mathbb{R}^n$.

PROOF. The second part of the statement is an immediate consequence of the first. The proof of the first part refines the classical proof in the $m = 0$ case. We write $c = (a, b)$ with $a \in \mathbb{Z}^m$, $b \in \mathbb{Z}^n$, so that $\gamma \cdot c = \alpha \cdot a + \beta \cdot b$ and $|c| = \sup(|a|, |b|)$. Let $B(\alpha, \tau)$ be the set of vectors β such that $|\gamma \cdot c| < |c|^{1-\tau}$ has an infinite number of solutions; we want to show that $B(\alpha, \tau)$ has measure 0. Let us refine $B(\alpha, \tau)$ into $B(\alpha, \tau, j, k)$ $(1 \leq j \leq m, 1 \leq k \leq n)$: by definition, the latter set contains the points β such that the inequality above has infinitely many solutions $c = (a, b)$ with $|a| = |a_j|$ and $|b| = |b_k|$. In other words over $B(\alpha, \tau, j, k)$, a_j (resp. b_k) is the largest component of a (resp. b). Now the—overlapping— mn sets $B(\alpha, \tau, j, k)$ cover $B(\alpha, \tau)$ and it is enough to prove that $B(\alpha, \tau, j, k)$ has Lebesgue measure 0. We may and will also restrict to the points β lying in the unit ball of $\mathbb{R}^n$.

We will now compute the asymptotic volume of the slices inside the unit ball of $\mathbb{R}^n$ corresponding to the fullfilment of the inequality $|\gamma \cdot c| < |c|^{1-\tau}$ for elements of $B(\alpha, \tau, j, k)$. First, given positive integers u and v, let $S(u, v)$ denote the "sphere" in $\mathbb{Z}^\ell$ of the elements $c = (a, b)$ with $|a| = u$, $|b| = v$. Let us restrict attention to $B(\alpha, \tau, j, k) \cap S(u, v)$ for some given pair (u, v). Because $\tau > \tau_1$ this set is empty if $v = 0$ and u is large enough (or equivalently $|c|$ is large enough) as we may and will assume, being in fact interested in asymptotic properties. So we may divide both sides of the inequality $|\alpha \cdot a + \beta \cdot b| < |c|^{1-\tau}$ by $v = |b|$. The set of β in the unit ball of $\mathbb{R}^n$ satisfying this inequality is now seen to have measure bounded above by $2|c|^{1-\tau}|b|^{-1} = 2w^{1-\tau}v^{-1}$ where $w = \sup(u, v)$. As u and v tend to infinity, the number of points in $S(u, v)$ grows like $u^{m-1}v^{n-1}$. So asymptotically in u, v, the number of elements in $B(\alpha, \tau, j, k) \cap S(u, v)$ grows (up to a multiplicative constant) like $w^{1-\tau}u^{m-1}v^{n-2} \leq w^{-\tau+\ell-2}$. Since we picked $\tau > \ell$, the double series of the measures of the sets $B(\alpha, \tau, j, k) \cap S(u, v)$ converges, as u and v run through the positive integers (recall that $w = \sup(u, v)$). An element of $B(\alpha, \tau, j, k)$ belongs to infinitely many sets $B(\alpha, \tau, j, k) \cap S(u, v)$ (in probabilistic terms, we are dealing with a "tail event"). Since the sum of the measures of these sets is finite, the measure of $B(\alpha, \tau, j, k)$ is zero by the Borel-Cantelli lemma. □

A.1.5. We finally briefly turn to the hyperbolic situation; we state and sketch the proof of the 1-hyperbolic version of Theorem A.1.1, because in particular that is what is needed to deal with Hamiltonian (∗). So the theorem below makes it possible to adapt some results about splitting to the case of vanishing torsion, by first detecting the appropriate tori and their invariant manifolds. It thus provides a natural bridge between isochronous and nonisochronous problems. Clearly many variants are again possible.

We repeat again the notation in this case, which is slightly different from that in the body of the paper as we partly follow that of the references we quote. We now have phase variables (I, ϕ, s, u) with $(I, \phi) \in \mathbb{R}^n \times \mathbb{T}^n$ and $(s, u) \in \mathbb{R}^2$ describing the hyperbolic part of the motion; coming back to Hamiltonian (*), they are given by integrating the (p, q) pendulum near the origin. We are focusing near an unperturbed n-torus $\mathcal{T}_0$ given by $I = s = u = 0$, with frequency ω, and attending stable (resp. unstable) invariant manifold $\mathcal{W}_0^+$ (resp. $\mathcal{W}_0^-$) with equation $I = u = 0$ (resp. $I = s = 0$). We denote by $\lambda > 0$ the positive eigenvalue corresponding to the hyperbolic directions. The statement now reads as follows:

THEOREM. *Consider the near-integrable Hamiltonian:*

$$H(I, \phi, s, u) = \omega I + \frac{1}{2}C_1 I_1^2 + \frac{1}{2}\tau C_2 I_2^2 + +\lambda s u + \varepsilon f(I_1, \phi, s, u) + \varepsilon \tau g(I, \phi, s, u),$$

where ε and τ are real parameters. Assume that the symmetric matrices C_1 and C_2 are invertible, that the functions f and g are defined and analytic near $I = s = u = 0$, and that the vector ω is Diophantine.

Then for $|\tau| \leq 1$ and $|\varepsilon| \leq \varepsilon_0$ with $\varepsilon_0 > 0$ independent of τ, there exists an invariant torus $\mathcal{T}_\varepsilon$ for H which is ε-close to $\mathcal{T}_0$ (again independently of τ) and on which the flow is conjugate to the rotation with frequency ω. Moreover $\mathcal{T}_\varepsilon$ has analytic invariant manifolds $\mathcal{W}_\varepsilon^\pm$ which are perturbations of $\mathcal{W}_0^\pm$ and on which the flow is integrable.

Again the conclusion is stated slightly informally and reads in fact just as the usual hyperbolic KAM statements. Here the setting is the same as in [Gr], except of course for the introduction of the torsion parameter τ. We refer to that paper for more details and precisions on the situation at hand. The proof follows [Gr] and we will be content with giving some indications. Other proofs of similar KAM-type theorems have been given in the partially hyperbolic situation, and one can also take advantage of further improvements, as are to be found *e.g.* in [Ni3]. The adaptation should be routine, although often tedious and quite cumbersome to write up in details. But one can indeed say that the real point of interest hitherto is embodied in Theorem A.1.1 and that the passage to the hyperbolic situation is innocuous. Returning to [Gr] and some indications on the proof, we first observe that we only need to worry about the persistence of the torus: Theorems 5 and 6 in [Gr] then guarantee the claims about the persistence of the invariant manifolds and the motion on them (which is a hyperbolic, rather than a symplectic phenomenon). Concerning the existence of the torus, the proof in [Gr] is quite similar to that in [K], although this is not completely immediate upon reading the statements. Indeed Theorem 4 of [Gr] states the persistence of the torus but does not explicitly imply the existence of a normalizing transformation converging in a neighbourhood of the torus in phase space. Yet this is is what one really gets, as stated in the italicized statement at the very end of Section 2 as an "alternate formulation of Theorem 4". Now in order to prove the theorem, it is enough to modify the first four pages of Section 3 in [Gr], exactly in the same way as we modified [K] for proving Theorem A.1.1. Using again Lie series or canonical transformations as in [Gr], one has to write the auxiliary Hamiltonian just as for χ above, introducing τ in

Formula (4) on p. 32 of [Gr]. One then gets a homological system which duplicates the system (5)-(10) in [Gr], and reduces to the one we wrote above for χ when forgetting about the hyperbolic variables s, u. The system and the estimates for its solutions are quite cumbersome to write down in details, but they completely parallel [Gr] with the same modifications as in the proof of Theorem A.1.1.

We note that the statement above does not immediately apply to Hamiltonian ($*$) as it describes the *a priori* unstable case, whereas in ($*$), $\lambda = \sqrt{\varepsilon}$ is small. So instead of [Gr] one has to resort to [T1], which describes how to adapt the scheme in [Gr] to the *a priori* stable case (yet a form of degeneracy), taking for instance μ polynomial with respect to ε in Hamiltonian ($*$). This further modification does not affect the reasoning about tori with vanishing or zero torsion, so that we do not detail it here.

BIBLIOGRAPHY

[A1] V.I.Arnold, Instability of dynamical systems with several degrees of freedom, Soviet Math. Doklady **5** (1964), 581–585.

[A2] V.I.Arnold, From Hilbert's superposition problem to dynamical systems, in *The Arnoldfest. Proceedings of a conference in Honour of V.I.Arnold for his Sixtieth Birthday*, E.Bierstone et al. eds., Fields Institute Communications Series **24**, AMS Publ., 1999.

[AG] V.I.Arnold, A.B.Givental, Symplectic Geometry, Encyclopedia of the Mathematical Sciences, Dynamical Systems **4**, Springer Verlag, 1990.

[AL] M.Audin, J.Lafontaine, *Holomorphic curves in symplectic geometry*, Prog. in Math **117**, Birkhäuser, 1994.

[Ba] D.Bambusi, Nekhoroshev theorem for small amplitude solutions in nonlinear Schrödinger equations, Math. Z. **130** (1999), 345–387.

[BG] D.Bambusi, A.Giorgilli, Exponential stability of states close to resonance in infinite dimensional Hamiltonian systems, J. Stat. Phys. **71** (1993), 569–606.

[BGG1] G.Benettin, L.Galgani, A.Giorgilli, A proof of Nekhoroshev's theorem for the stability times in nearly integrable Hamiltonian systems, Celestial Mechanics **37** (1985), 1–25.

[BGG2] G.Benettin, L.Galgani, A.Giorgilli, Boltzmann's ultraviolet cutoff and Nekhoroshev's theorem, Nature **311** (1984), 444–445.

[BGGS] G.Benettin, L.Galgani, A.Giorgilli, J.-M.Strelcyn, A proof of Kolmogorov's theorem on invariant tori etc., Il Nuovo Cimento **79 B** (1984), 201–223.

[B] P.Bernard, Perturbation d'un Hamiltonien partiellement hyperbolique, C. R. Acad. Sci. Paris Sér. I Math. **323** (1996), 189–194.

[Be1] U.Bessi, An approach to Arnold diffusion through the calculus of variations, Nonlinear Analysis T.M.A. **26** (1996), 1115–1135.

[Be2] U.Bessi, Arnold's example with three rotators, Nonlinearity **10** (1997), 763–781.

[Be3] U.Bessi, An analytic counterexample to the KAM theorem, Ergod. Th. and Dynam. Sys. 20 (2000), 317–333.

[Be4] U.Bessi, A λ-lemma for transition tori, Preprint, 1999.

[Bol1] S.V.Bolotin, Homoclinic orbits to invariant tori of Hamiltonian systems, AMS Transl. **168** (1995), 21–90.

[Bol2] S.V.Bolotin, Homoclinic trajectories of invariant sets of Hamiltonian systems, Nonlinear Diff. Eq. and Appl. **4** (1997), 359–389.

[BT] S.V.Bolotin, D.V.Treschev, Unbounded growth of energy in nonautonomous Hamiltonian systems, Nonlinearity **12** (1999), 365–388.

[Bo] J.-B.Bost, Tores invariants des systèmes dynamiques hamiltoniens, Astérisque **133–134** (1986), 113–157.

[Ch] M.Chaperon, Some results on stable manifolds, C. R. Acad. Sci. Paris Sér. I Math. **333** (2001), 119-124.

[CG] L.Chierchia and G.Gallavotti, Drift and diffusion in phase space, Annales de l'IHP, Section Physique Théorique **60** (1994), 1–144; see also the Erratum in vol. **68** (1998), 135.

[C] B.V.Chirikov, A universal instability of many-dimensional oscillator systems, Phys. Reports **52** (1979), 263–379.

[CV1] B.V.Chirikov and V.V.Vecheslavov, KAM Integrability, in *Analysis, et cetera*, P.Rabinowitz and E.Zehnder eds., 219–236, Academic Press, 1990.

[CV2] B.V.Chirikov and V.V.Vecheslavov, How fast is Arnold diffusion?, in Proceedings of the ICFA Beam Dynamics Workshop, Novossibirsk, 1989, 39–58.

[CS] K.Cieliebak and É.Séré, Pseudo-holomorphic curves and multiplicity of homoclinic orbits, Duke Math. J. **77** (1995), 483–518.

[Cr1] J.Cresson, Symbolic dynamics and Arnold diffusion, Preprint Université de Besançon, 1998.

[Cr2] J.Cresson, Conjecture de Chirikov et optimalité des exposants de stabilité du théorème de Nekhorochev, Preprint Université de Besançon, 1998.

[DGJS] A.Delshams, V.Gelfreich, À.Jorba and T.M.Seara, Exponentially small splitting of separatrices under fast quasi-periodic forcing, Commun. Math. Phys. **189** (1997), 35–71.

[DG1] A.Delshams and P.Gutiérrez, Homoclinic orbits to invariant tori in Hamiltonian systems, in *Multiple-Time-Scale Dynamical Systems*, C.Jones et al. eds., IMA Vol. in Math. and its Appl., Springer, 1998.

[DG2] A.Delshams and P.Gutiérrez, Splitting potential and the Poincaré-Melnikov method for whiskered tori in Hamiltonian systems, J. Nonlinear Sci. **10** (2000), 433–476.

[DS] A.Delshams and T.M.Seara, An asymptotic expression for the splitting of separatrices of the rapidly forced pendulum, Commun. Math. Phys. **150** (1992), 433–463.

[El] L.H.Eliasson, Biasymptotic solutions of perturbed integrable Hamiltonian systems, Bull. Soc. Bras. Mat. **25** (1994), 57–76.

[EKS] J.A.Ellison, M.Kummer and A.W.Sáenz, Transcendentally small transversality in the rapidly forced pendulum, J. Dyn. Diff. Eq. **5** (1993), 241–277.

[F] F.Fassò, Lie series method for vector fields and Hamiltonian perturbation theory, ZAMP **41** (1990), 843–864.

[Fa] A.Fathi, Orbites hétéroclines et ensembles de Peierls, C. R. Acad. Sci. Paris Sér. I Math. **326** (1998), 1213–1216.

[Ga1] G.Gallavotti, Twistless KAM tori, Commun. Math. Phys. **164** (1994), 145–156.

[Ga2] G.Gallavotti, Twistless KAM tori, quasi flat homoclinic intersections, and other cancellations in the perturbation series of certain completely integrable systems; a review, Rev. Math. Phys. **6** (1994), 343–411.

[Ga-Ge] G.Gallavotti and G.Gentile, Majorant series convergence for twistless KAM tori, Ergod. Th. and Dynam. Sys. **15** (1995), 857–869.

[GGM1] G.Gallavotti, G.Gentile and V.Mastropietro, Separatrix splitting for systems with three timescales, Commun. Math. Phys. **202** (1999), 197–236.

[GGM2] G.Gallavotti, G.Gentile and V.Mastropietro, Melnikov's approximation dominance. Some examples, Rev. Math. Phys. **11** (1999), 451–461.

[G1] V.G.Gelfreich, Reference systems for splitting of separatrices, Nonlinearity **10** (1997), 175–193.

[G2] V.G.Gelfreich, A proof of the exponentially small transversality of the separatrices for the standard map, Commun. Math. Phys. **201** (1999), 155–216.

[GLS] V.Gelfreich, V.F.Lazutkin and N.V.Svanidze, A refined formula for the separatrix splitting for the standard map, Phys. D **71** (1994), 82–101.

[Gr] S.M.Graff, On the conservation of hyperbolic invariant tori for Hamiltonian systems, J. of Diff. Eq. **15** (1974), 1–69.

[GS] V.Guillemin, S.Sternberg, *Symplectic techniques in physics*, Cambridge University Press, 1984.

[HK] B.Hasselblatt, A.Katok, *Introduction to the modern theory of dynamical systems*, Cambridge University Press (1995).

[H1] M.Herman On the dynamics on Lagrangian tori invariant by symplectic diffeomorphisms, in *Progress in variational methods etc.*, 92–112, Longman Sci. Tech., Harlow, 1992.

[H2] M.Herman, Inégalités a priori pour des tores Lagrangiens invariants par des difféomorphismes symplectiques, Publ. Math. IHES **70** (1989), 47–101.

[HPS] M.W.Hirsch, C.C.Pugh, M. Shub, *Invariant manifolds*, LN in Math. **583**, Springer Verlag, 1977.

[HMS] P.Holmes, J.Marsden and J.Scheurle, Exponentially small splittings of separatrices with applications to KAM theory and degenerate bifurcations, Contemporary Math. **81** (1988), 213–244.

[K] A.N.Kolmogorov, On the preservation of conditionally periodic motions, Doklady AN **98** (1954), 527–530. Reprinted in *Selected works of A.N.Kolmogorov. Volume I: Mathematics and Mechanics*, V.M.Tikhomirov ed., Kluwer Acad. Publ., 1991.

[La1] V.F.Lazutkin, Splitting of the separatrices of Chirikov's standard map, Preprint VINITI No. 6372-84 (in Russian), 1984.

[La2] V.F.Lazutkin, An analytic integral along the separatrix of the semistandard map, St Petersburg Math. J. **4** (1993), 721–748.

[LST] V.F.Lazutkin, I.G.Schahmannski and M.B.Tabanov, Splitting of separatrices for standard and semistandard mappings, Phys. D **40** (1989), 235–248.

[Li-Ma] P.Lieberman and C.-M.Marle, *Symplectic geometry and analytical mechanics*, D. Reidel Pub., 1987.

[L1] P.Lochak, Effective speed of Arnold diffusion and small denominators, Physics Letters A **143** (1990), 39–42.

[L2] P.Lochak, Canonical perturbation theory via simultaneous approximation, Russian Math. Surveys **47** (1992), 57–133.

[L3] P.Lochak, Hamiltonian perturbation theory: periodic orbits, resonances and intermittency, Nonlinearity **6** (1993), 885–904.

[L4] P.Lochak, Arnold diffusion: a compendium of remarks and questions, in *Hamiltonian systems with three or more degrees of freedom*, C.Simó ed., Kluwer Acad. Publ., 1999, 168–183.

[L5] P.Lochak, Stability of Hamiltonian systems over exponentially long times: the near-linear case, in *Hamiltonian Dynamical Sytems: History, Theory and Applications*, IMA Volumes in Math. and Appl. **63**, Springer Verlag, H.S. Dumas, K.R. Mayer and D.S. Schmidt eds., 1995, 221–229.

[L6] P.Lochak, Tores invariants à torsion évanescente, C. R. Acad. Sci. Paris Sér. I Math. **327** (1998), 833–836.

[LM] P.Lochak and C.Meunier, *Multiphase averaging for classical systems*, Appl. Math. Sciences Series **72**, Springer Verlag, 1988.

[Mar1] J.-P.Marco, Transition le long des chaînes de tores invariants pour les systèmes hamiltoniens analytiques, Ann. Inst. Poincaré **64** (1996), 205–252.

[Mar2] J.-P.Marco, Dynamics in the vicinity of double resonances, Proceedings of the 4th Catalan Days of Applied Math., C.Garcia *et al.* eds., 1998.

[MDS] D.Mc Duff, D.Salamon, *Introduction to symplectic topology*, Oxford University Press, 1995.

[N]   N.N.Nekhoroshev, An exponential estimate for the time of stability of nearly integrable Hamiltonian systems, Russian Math. Surveys **32** (1977), 1–65.

[Nic] A.J.Nicas, Classifying pairs of Lagrangians in a hermitian vector space, Topology and its Applications **42** (1991), 71-81.

[Ni1] L.Niederman, Stability over exponentially long times in the planetary problem, Nonlinearity **9** (1996), 1703–1751.

[Ni2] L.Niederman, Nonlinear stability around an elliptic equilibrium point in a Hamiltonian system, Nonlinearity **11** (1998), 1465–1479.

[Ni3] L.Niederman, Dynamics around simple resonant tori in nearly integrable Hamiltonian sytems, J. of Diff. Eq. **161** (2000), 1–41.

[P]   H.Poincaré, Méthodes nouvelles de la mécanique céleste, 3 Volumes, Gauthier Villars, 1892, 1893, 1899.

[Pö1] J.Pöschel, Nekhoroshev estimates for quasi-convex Hamiltonian sytems, Math. Z. **213** (1993), 187–216.

[Pö2] J.Pöschel, On Nekhoroshev estimates at an elliptic equilibrium, Internat. Math. Res. Notices **4** (1999), 203–215.

[Pö3] J.Pöschel, Nekhoroshev estimates for a nonlinear Schrödinger equation and a theorem by Bambusi, Nonlinearity **12** (1999), 1587–1600.

[RS]   M.Reed and B.Simon, *Methods of Modern Mathematical Physics*, 4 volumes, Academic Press, 1978.

[RW1] M.Rudnev and S.Wiggins, Existence of exponentially small separatrix splittings and homoclinic connections between whiskered tori in weakly hyperbolic near-integrable Hamiltonian systems, Phys. D **114** (1998), 3–80; see also the erratum in Phys. D **145** (2000), 349–354.

[RW2] M.Rudnev and S.Wiggins, On the dominant Fourier modes in the series associated with separatrix splitting for an a priori stable, three degree-of-freedom Hamiltonian system, in *The Arnoldfest. Proceedings of a conference in Honour of V.I.Arnold for his Sixtieth Birthday*, E.Bierstone et al. eds., Fields Institute Communications Series **24**, AMS Publ., 1999.

[RW3] M.Rudnev and S.Wiggins, On a homoclinic splitting problem, Regul. Chaotic Dyn. **5** (2000), 227–242.

[R1]   H.Rüssmann, Nondegeneracy in the perturbation theory of integrable dynamical systems, in *Stochastics, algebra and analysis in classical and quantum systems*, 211–223, Kluwer Acad. Publ., 1990.

[R2]   H.Rüssmann, On the frequencies of quasi-periodic solutions of analytic nearly integrable Hamiltonian systems, in *Seminar on Dynamical Systems*, Progr. Nonlinear Diff. Eq. Appl. **12**, Birkhäuser, 1994.

[R3]   H.Rüssmann, Invariant tori in non-degenerate nearly integrable Hamiltonian systems, Regul. Chaotic Dyn. **6** (2001), 119–204.

[Sa1] D.Sauzin, Résurgence paramétrique et exponentielle petitesse de l'écart des séparatrices du pendule rapidement forcé, Annales de l'Institut Fourier **45** (1995), 453–511.

[Sa2] D.Sauzin, A new method for measuring the splitting of invariant manifolds, Ann. Scient. Ec. Norm. Sup. **34** (2001), 159–221.

[ST]   L.P.Shilnikov, D.V.Turaev, Hamiltonian systems with homoclinic saddle curves, Soviet Math. Dokl. **39** (1989), 165–168.

[S] C.Simó, Averaging under fast quasiperiodic forcing, in *Hamiltonian mechanics: integrability and chaotic behaviour*, vol. **331** of NATO Adv. Sci. Inst. Ser. B Phys., Plenum, 1994, 13–34.

[St] K.H.Strobel, Multibump solutions for a class of periodic Hamiltonian systems, PhD Thesis, University of Wisconsin, 1994.

[T1] D.V.Treshchëv, A mechanism for the destruction of resonance tori in Hamiltonian systems, Math. USSR-Sbornik **68** (1991), 181–203.

[T2] D.V.Treschev, Hyperbolic tori and asymptotic surfaces in Hamiltonian systems, Russian J. Math. Phys. **2** (1994), 93–110.

[T3] D.V.Treshchëv, Splitting of separatrices from the point of view of symplectic geometry, Math. Notes **61** (1997), 744–757.

[W] S.Wiggins, *Normally hyperbolic invariant manifolds in dynamical systems*, Appl. Math. Sciences Series **105**, Springer Verlag, 1994.

[Y1] J.-C.Yoccoz, Travaux de Herman sur les tores invariants, Séminaire Bourbaki 754, Astérisque **201–203** (1992), 143–165.

[Y2] J.-C.Yoccoz, Introduction to hyperbolic dynamics, in *Real and complex dynamical systems*, 265–291, Kluwer Acad. Publ., 1995.

Editorial Information

To be published in the *Memoirs*, a paper must be correct, new, nontrivial, and significant. Further, it must be well written and of interest to a substantial number of mathematicians. Piecemeal results, such as an inconclusive step toward an unproved major theorem or a minor variation on a known result, are in general not acceptable for publication. Papers appearing in *Memoirs* are generally longer than those appearing in *Transactions*, which shares the same editorial committee.

As of February 1, 2003, the backlog for this journal was approximately 3 volumes. This estimate is the result of dividing the number of manuscripts for this journal in the Providence office that have not yet gone to the printer on the above date by the average number of monographs per volume over the previous twelve months, reduced by the number of volumes published in four months (the time necessary for preparing a volume for the printer). (There are 6 volumes per year, each containing at least 4 numbers.)

A Consent to Publish and Copyright Agreement is required before a paper will be published in the *Memoirs*. After a paper is accepted for publication, the Providence office will send a Consent to Publish and Copyright Agreement to all authors of the paper. By submitting a paper to the *Memoirs*, authors certify that the results have not been submitted to nor are they under consideration for publication by another journal, conference proceedings, or similar publication.

Information for Authors

Memoirs are printed from camera copy fully prepared by the author. This means that the finished book will look exactly like the copy submitted.

The paper must contain a *descriptive title* and an *abstract* that summarizes the article in language suitable for workers in the general field (algebra, analysis, etc.). The *descriptive title* should be short, but informative; useless or vague phrases such as "some remarks about" or "concerning" should be avoided. The *abstract* should be at least one complete sentence, and at most 300 words. Included with the footnotes to the paper should be the 2000 *Mathematics Subject Classification* representing the primary and secondary subjects of the article. The classifications are accessible from www.ams.org/msc/. The list of classifications is also available in print starting with the 1999 annual index of *Mathematical Reviews*. The Mathematics Subject Classification footnote may be followed by a list of *key words and phrases* describing the subject matter of the article and taken from it. Journal abbreviations used in bibliographies are listed in the latest *Mathematical Reviews* annual index. The series abbreviations are also accessible from www.ams.org/publications/. To help in preparing and verifying references, the AMS offers MR Lookup, a Reference Tool for Linking, at www.ams.org/mrlookup/. When the manuscript is submitted, authors should supply the editor with electronic addresses if available. These will be printed after the postal address at the end of the article.

Electronically prepared manuscripts. The AMS encourages electronically prepared manuscripts, with a strong preference for $\mathcal{AMS}$-LaTeX. To this end, the Society has prepared $\mathcal{AMS}$-LaTeX author packages for each AMS publication. Author packages include instructions for preparing electronic manuscripts, the *AMS Author Handbook*, samples, and a style file that generates the particular design specifications of that publication series. Though $\mathcal{AMS}$-LaTeX is the highly preferred format of TeX, author packages are also available in $\mathcal{AMS}$-TeX.

Authors may retrieve an author package from e-MATH starting from `www.ams.org/tex/` or via FTP to `ftp.ams.org` (login as `anonymous`, enter username as password, and type `cd pub/author-info`). The *AMS Author Handbook* and the *Instruction Manual* are available in PDF format following the author packages link from `www.ams.org/tex/`. The author package can be obtained free of charge by sending email to `pub@ams.org` (Internet) or from the Publication Division, American Mathematical Society, P.O. Box 6248, Providence, RI 02940-6248. When requesting an author package, please specify $\mathcal{AMS}$-LaTeX or $\mathcal{AMS}$-TeX, Macintosh or IBM (3.5) format, and the publication in which your paper will appear. Please be sure to include your complete mailing address.

Sending electronic files. After acceptance, the source file(s) should be sent to the Providence office (this includes any TeX source file, any graphics files, and the DVI or PostScript file).

Before sending the source file, be sure you have proofread your paper carefully. The files you send must be the EXACT files used to generate the proof copy that was accepted for publication. For all publications, authors are required to send a printed copy of their paper, which exactly matches the copy approved for publication, along with any graphics that will appear in the paper.

TeX files may be submitted by email, FTP, or on diskette. The DVI file(s) and PostScript files should be submitted only by FTP or on diskette unless they are encoded properly to submit through email. (DVI files are binary and PostScript files tend to be very large.)

Electronically prepared manuscripts can be sent via email to `pub-submit@ams.org` (Internet). The subject line of the message should include the publication code to identify it as a Memoir. TeX source files, DVI files, and PostScript files can be transferred over the Internet by FTP to the Internet node `e-math.ams.org` (130.44.1.100).

Electronic graphics. Comprehensive instructions on preparing graphics are available at `www.ams.org/jourhtml/graphics.html`. A few of the major requirements are given here.

Submit files for graphics as EPS (Encapsulated PostScript) files. This includes graphics originated via a graphics application as well as scanned photographs or other computer-generated images. If this is not possible, TIFF files are acceptable as long as they can be opened in Adobe Photoshop or Illustrator. No matter what method was used to produce the graphic, it is necessary to provide a paper copy to the AMS.

Authors using graphics packages for the creation of electronic art should also avoid the use of any lines thinner than 0.5 points in width. Many graphics packages allow the user to specify a "hairline" for a very thin line. Hairlines often look acceptable when proofed on a typical laser printer. However, when produced on a high-resolution laser imagesetter, hairlines become nearly invisible and will be lost entirely in the final printing process.

Screens should be set to values between 15% and 85%. Screens which fall outside of this range are too light or too dark to print correctly. Variations of screens within a graphic should be no less than 10%.

Inquiries. Any inquiries concerning a paper that has been accepted for publication should be sent directly to the Electronic Prepress Department, American Mathematical Society, P. O. Box 6248, Providence, RI 02940-6248.

Editors

This journal is designed particularly for long research papers, normally at least 80 pages in length, and groups of cognate papers in pure and applied mathematics. Papers intended for publication in the *Memoirs* should be addressed to one of the following editors. In principle the Memoirs welcomes electronic submissions, and some of the editors, those whose names appear below with an asterisk (*), have indicated that they prefer them. However, editors reserve the right to request hard copies after papers have been submitted electronically. Authors are advised to make preliminary email inquiries to editors about whether they are likely to be able to handle submissions in a particular electronic form.

Algebra to KAREN E. SMITH, Department of Mathematics, University of Michigan, 525 University, Suite 2832, Ann Arbor, MI 48109-1109; email: `kesmith@lsa.umich.edu`

Algebraic geometry to DAN ABRAMOVICH, Department of Mathematics, Boston University, 111 Cummington Street, Boston, MA 02215; e-mail: `abrmovic@bu.edu`

Algebraic topology and cohomology of groups to STEWART PRIDDY, Department of Mathematics, Northwestern University, 2033 Sheridan Road, Evanston, IL 60208-2730; email: `priddy@math.nwu.edu`

Combinatorics and Lie theory to SERGEY FOMIN, Department of Mathematics, University of Michigan, Ann Arbor, Michigan 48109-1109; email: `fomin@umich.edu`

Complex analysis and complex geometry to DUONG H. PHONG, Department of Mathematics, Columbia University, 2990 Broadway, New York, NY 10027-0029; email: `phong@math.columbia.edu`

*****Differential geometry and global analysis** to LISA C. JEFFREY, Department of Mathematics, University of Toronto, 100 St. George St., Toronto, ON Canada M5S 3G3; email: `jeffrey@math.toronto.edu`

Dynamical systems and ergodic theory to ROBERT F. WILLIAMS, Department of Mathematics, University of Texas, Austin, Texas 78712-1082; email: `bob@math.utexas.edu`

*****Geometric analysis** to TOBIAS COLDING, Courant Institute, New York University, 251 Mercer Street, New York, NY 10012; email: `colding@cims.nyu.edu`

Geometric topology, knot theory and hyperbolic geometry to ABIGAIL A. THOMPSON, Department of Mathematics, University of California, Davis, Davis, CA 95616-5224; email: `thompson@math.ucdavis.edu`

Harmonic analysis, representation theory, and Lie theory to ROBERT J. STANTON, Department of Mathematics, The Ohio State University, 231 West 18th Avenue, Columbus, OH 43210-1174; email: `stanton@math.ohio-state.edu`

*****Logic** to THEODORE SLAMAN, Department of Mathematics, University of California, Berkeley, CA 94720-3840; email: `slaman@math.berkeley.edu`

Number theory to HAROLD G. DIAMOND, Department of Mathematics, University of Illinois, 1409 W. Green St., Urbana, IL 61801-2917; email: `diamond@math.uiuc.edu`

*****Ordinary differential equations, and applied mathematics** to PETER W. BATES, Department of Mathematics, Michigan State University, East Lansing, MI 48824-1027; email: `peter@math.msu.edu`

*****Partial differential equations** to PATRICIA E. BAUMAN, Department of Mathematics, Purdue University, West Lafayette, IN 47907-1395' email: `bauman@math.purdue.edu`

*****Probability and statistics** to KRZYSZTOF BURDZY, Department of Mathematics, University of Washington, Box 354350, Seattle, Washington 98195-4350; email: `burdzy@math.washington.edu`

Real analysis and partial differential equations to DANIEL TATARU, Department of Mathematics, University of California, Berkeley, Berkeley, CA 94720; email: `tataru@math.berkeley.edu`

All other communications to the editors should be addressed to the Managing Editor, WILLIAM BECKNER, Department of Mathematics, University of Texas, Austin, TX 78712-1082; email: `beckner@math.utexas.edu`.

Titles in This Series

775 **P. Lochak, J.-P. Marco, and D. Sauzin,** On the splitting of invariant manifolds in multidimensional near-integrable Hamiltonian systems, 2003

774 **Kai A. Behrend,** Derived ℓ-adic categories for algebraic stacks, 2003

773 **Robert M. Guralnick, Peter Müller, and Jan Saxl,** The rational function analogue of a question of Schur and exceptionality of permutation representations, 2003

772 **Katrina Barron,** The moduli space of $N = 1$ superspheres with tubes and the sewing operation, 2003

771 **Shigenori Matsumoto,** Affine flows on 3-manifolds, 2003

770 **W. N. Everitt and L. Markus,** Elliptic partial differential operators and symplectic algebra, 2003

769 **Jie Wu,** Homotopy theory of the suspensions of the projective plane, 2003

768 **R. Höpfner and E. Löcherbach,** Limit theorems for null recurrent Markov processes, 2003

767 **Po Hu,** S-modules in the category of schemes, 2003

766 **Su Gao and Alexander S. Kechris,** On the classification of Polish metric spaces up to isometry, 2003

765 **Robert Bieri and Ross Geoghegan,** Connectivity properties of group actions on non-positively curved spaces, 2003

764 **J. Spandaw,** Noether-Lefschetz problems for degeneracy loci, 2003

763 **Yasuyuki Kachi and Eiichi Sato,** Segre's reflexivity and an inductive characterization os hyperquadrics, 2002

762 **Leiba Rodman, Ilya M. Spitkovsky, and Hugo Woerdeman,** Abstract band method via factorization, positive and band extensions of multivariable almost periodic matrix functions, and spectral estimation, 2002

761 **Oliver Druet and Emmanuel Hebey,** The AB program in geometric analysis : Sharp Sobolev inequalities and related problems, 2002

760 **Markus Banagl,** Extending intersection homology type invarients to non-Witt spaces, 2002

759 **Donald M. Davis,** From representation theory to homotopy groups, 2002

758 **Alan Forrest, John Hunton, and Johannes Kellendonk,** Topological invariants for projection method patterns, 2002

757 **Douglas Bowman,** q-difference operators, orthogonal polynomials, and symmetric expansions, 2002

756 **José Ignacio Cogolludo-Agustín,** Topological invariants of the complement to arrangements of rational plane curves, 2002

755 **M. A. Mandell and J. P. May,** Equivariant orthogonal spectra and S-modules, 2002

754 **Edward L. Green, Idun Reiten, and Øyvind Solberg,** Dualities on generalized Koszul algebras, 2002

753 **Daniel Panazzolo,** Desingularization of nilpotent singularities in families of planar vector fields, 2002

752 **Linus Kramer,** Homogeneous spaces, Tits buildings, and isoparametric hypersurfaces, 2002

751 **Bruce Allison, Georgia Benkart, and Yun Gao,** Lie algebras graded by the root systems BC_r, $r \geq 2$, 2002

750 **Masaki Izumi and Hideki Kosaki,** Kac algebras arising from composition of subfactors: General theory and classification, 2002

749 **Nanhua Xi,** The based ring of two-sided cells of affine Weyl groups of type $\widetilde{A}_{n-1}$, 2002

748 **Jürgen Ritter and Alfred Weiss,** The lifted root number conjecture and Iwasawa theory, 2002

TITLES IN THIS SERIES

747 **Armand Borel, Robert Friedman, and John W. Morgan,** Almost commuting elements in compact Lie groups, 2002

746 **Peter Niemann,** Some generalized Kac-Moody algebras with known root multiplicities, 2002

745 **Mikhail A. Lifshits and Werner Linde,** Approximation and entropy numbers of Volterra operators with application to Brownian motion, 2002

744 **Roger Chalkley,** Basic global relative invariants for homogeneous linear differential equations, 2002

743 **Heng Sun,** Spectral decomposition of a covering of $GL(r)$: the Borel case, 2002

742 **J. E. Gilbert, Y. S. Han, J. A. Hogan, J. D. Lakey, D. Weiland, and G. Weiss,** Smooth molecular functions and singular integral operators, 2002

741 **Francisco Santos,** Triangulations of oriented matroids, 2002

740 **Rick Durrett,** Mutual invadability implies coexistence in spatial models, 2002

739 **Georgios K. Alexopoulos,** Sub-Laplacians with drift on Lie groups of polynomial volume growth, 2002

738 **Yasuro Gon,** Generalized Whittaker functions on $SU(2,2)$ with respect to the Siegel parabolic subgroup, 2002

737 **Arjen Doelman, Robert A. Gardner, and Tasso J. Kaper,** A stability index analysis of 1-D patterns of the Gray-Scott model, 2002

736 **Wojciech Chachólski and Jérôme Scherer,** Homotopy theory of diagrams, 2002

735 **Martina Brück, Xi Du, Joonsang Park, and Chuu-Lian Terng,** The submanifold geometries associated to Grassmannian systems, 2002

734 **Michel Van den Bergh,** Blowing up of non-commutative smooth surfaces, 2001

733 **Milé Krajčevski,** Tilings of the plane, hyperbolic groups and small cancellation conditions, 2001

732 **Jan O. Kleppe, Juan C. Migliore, Rosa Miró-Roig, Uwe Nagel, and Chris Peterson,** Gorenstein liaison, complete intersection liaison invariants and unobstructedness, 2001

731 **Jesús Bastero, Mario Milman, and Francisco J. Ruiz,** On the connection between weighted norm inequalities, commutators and real interpolation, 2001

730 **Suhyoung Choi,** The decomposition and classification of radiant affine 3-manifolds, 2001

729 **Michael Grosser, Eva Farkas, Michael Kunzinger, and Roland Steinbauer,** On the foundations of nonlinear generalized functions I and II, 2001

728 **Laura Smithies,** Equivariant analytic localization of group representations, 2001

727 **Anthony D. Blaom,** A geometric setting for Hamiltonian perturbation theory, 2001

726 **Victor L. Shapiro,** Singular quasilinearity and higher eigenvalues, 2001

725 **Jean-Pierre Rosay and Edgar Lee Stout,** Strong boundary values, analytic functionals, and nonlinear Paley-Wiener theory, 2001

724 **Lisa Carbone,** Non-uniform lattices on uniform trees, 2001

723 **Deborah M. King and John B. Strantzen,** Maximum entropy of cycles of even period, 2001

722 **Hernán Cendra, Jerrold E. Marsden, and Tudor S. Ratiu,** Lagrangian reduction by stages, 2001

721 **Ingrid C. Bauer,** Surfaces with $K^2 = 7$ and $p_g = 4$, 2001

720 **Palle E. T. Jorgensen,** Ruelle operators: Functions which are harmonic with respect to a transfer operator, 2001

For a complete list of titles in this series, visit the
AMS Bookstore at **www.ams.org/bookstore/**.